T0155898

Safety Accidents in Risky Industries

Safety Accidents in Risky Industries

Black Swans, Gray Rhinos and Other Adverse Events

Sasho Andonov

CRC Press is an imprint of the
Taylor & Francis Group, an informa business

First edition published 2022
by CRC Press
6000 Broken Sound Parkway NW, Suite 300, Boca Raton, FL 33487-2742

and by CRC Press
2 Park Square, Milton Park, Abingdon, Oxon, OX14 4RN

First edition published by CRC Press 2022

CRC Press is an imprint of Taylor & Francis Group, LLC

ISBN: 9781032136479 (hbk)
ISBN: 9781032136493 (pbk)
ISBN: 9781003230298 (ebk)

DOI: 10.1201/9781003230298

Typeset in Times
by codeMantra

"To think that nothing bad could happen
to you, you need to be stupid!

To think that you can survive everything bad which
can happen to you, you need to be prepared!"

(Said by a very wise guy...)

Contents

Preface

Working at the Military Technological College (MTC) was a great experience for me because being around students was an incredible joy (which I had just discovered)!

After finishing the classes, I had enough time to deal with other things. During the afternoons and at the weekends, after discovering the beauties of Oman, I could dedicate myself to the ideas coming to my mind very often.

I read a lot!

Even today, "when I am too old for rock'n'roll and too young to die", I still have a passion for reading and learning, just as when I was a child…

Having in mind that my family was not with me in Muscat, I was using every opportunity to read books from the MTC Library or books downloaded from Internet. When I got the idea for this book, I was writing my second book.[1] Looking for available literature for the Bowtie Methodology, I went through plenty of books, papers, and articles available on the Internet and in the MTC library. Searching the Internet, my attention was drawn by a very interesting book…

It was the Master thesis of Adesanya Adeleke Oluwole. The name of the thesis was "Strengths and Weaknesses of Anticipatory Failure Determination (AFD) in Identifying Black Swan Type of Events" and it was submitted to the Faculty of Science and Technology at the University of Stavanger (Norway) in June 2014. This thesis did not contribute at all to my second book, but the term "Black Swans" occupied my attention and I tried to investigate "what is going on" on the Internet. That is how I found the information about the book *The Black Swan: The Impact of Highly Improbable* by Nassim Nicholas Taleb (NNT).

I bought NNT's book and I read it. I was delighted by the book and although NNT spoke about the Black Swan events (BSe) in the economy and the financial world (especially in the Stock Exchange), I immediately started to think about the existence of the Black Swan events (BSe) in the Risky Industries.[2] The idea for the new research and, the possibility of a new book, just popped up in my head…

That was also the time when I tried to find a reason why the Functional Safety is in stagnation in the Risky Industries…

You will be shocked by this statement, but starting from the year 2000 (approximately) when the Regulatory requirement for implementing a Safety Management System[3] (SMS) in aviation happened, I could notice that there was no progress in the Safety area as it happened in the Quality area. Even the implementation and the maintenance of the SMS in many companies from the Risky Industries

---

[1] "Bowtie Methodology: A Guide for Practitioners", issued by CRC Press in USA in October 2017.

[2] The Risky Industries are industries which could endanger humans, assets and environment in a huge scale by their operations, activities, processes, services and products. Such industries are: aviation, railway, nuclear, chemical, pharmaceutical, medical, etc.

[3] There are two types of safety in Industry: Occupational Health & Safety (OHS) and Functional Safety. I am speaking about introduction of Functional Safety in aviation which happened approximately in that time. Anyway, this is a book which applies to Functional Safety in the Risky Industries, although I will use in the book only a general term: Safety.

is still more bureaucratic than fundamental, as it shall be. In comparison, since the Quality Management System (QMS) was established as a systematic and structured way to deal with the quality in industry, there was tremendous development of new tools and new methods for use in areas of Quality Control and Quality Assurance (QC/QA). Today, in the Quality area, we have Six Sigma, the methodology which is the "Ferrari" between the quality management methodologies, but in the safety area nothing moved so significantly.

Some would say that there is Safety-II, the new approach dealing with "what is going right", but this is actually a new name for something which I call Quality-I. I have written a book regarding Safety-I and Safety-II and, there, the connection of Safety-II with the Quality Management System is explained, so there is no reason to explain it again here. Anyway, there is nothing revolutionary in the implementation of the Safety-II: It is just a concept developed by the safety guys who did/do not understand the quality.

That was also the time when I was looking for some directions where safety may progress and I found the Black Swans as an area worth investigating and checking, as this can really contribute to improving the safety in the Risky Industries.

Preparing this book, I found plenty of papers about the Black Swan events (BSe), but there were no papers dealing with the Black Swan events in the Risky Industries the way it is presented in this book. That is the reason that, I do believe, this book will contribute to increasing the safety in the Risky Industries by understanding all these events.

Later, during my research about available literature regarding the book, I found also another book *The Gray Rhino* by Michele Wucker. This book could not be ignored by me: I bought it and I read it. This is a book which deals with a slightly different type of adverse events (incidents and accidents), named "Gray Rhino" events (GRe) in the book. These are (mostly) bad catastrophic events in the economy, the politics, and the business fields, or in other words: In the big part of our ordinary lives. Immediately, I realized that, if I would like to provide to the safety community a holistic book about new ideas and directions for the safety, the book for Black Swan events should include also the Gray Rhinos events (GRe).

After further investigation into the areas of the BSe and the GRe, I have noticed that there are plenty of such "safety animals" that are mostly metaphors for different types of adverse events. The "fashion" which started with NNT's Black Swan was used by many other authors to point to different characteristics of other adverse events. In this book, I will use these metaphoric names to explain the influence of all these "safety animals" in providing the safety of the operations in the Risky Industries. It does not mean that Occupational Health & Safety (OHS) is immune to these "safety animals", but the situation there is very well regulated and what is applying for the Functional Safety will also apply to the OHS.

At the time when I started to write this book, I was not sure what the research about these few metaphors would produce. But today, I am convinced that they should be addressed and our mental adjustments regarding the Functional Safety must be updated accordingly. I must say that the BSe and GRe may look only as pretty different aspects of the accidents and the incidents in our lives, but their connection to uncertainty and the unpredictability is so evident, so the future development of safety must take care of them.

I hope that this book is just the beginning regarding effects of these "safety animals" in the Risky Industries. This is the reason that it is more scientifically based, but with characteristic and intention to be a popular book. Clearly, this is a book about metaphors and it is dedicated to the safety theoreticians and the safety practitioners. I try to give an analysis and possible remedies, as well as my opinion for these events. I hope that all "pros" and "cons" presented in this book will occupy reader's attention, and I hope this book will provoke more research regarding the "safety animals" in the future.

Sasho Andonov
Skopje, 11/09/2021

Author

Sasho Andonov is Graduate Engineer of Electronics and Telecommunications, and he has a Master's Degree in Metrology and Quality Management. He has 30 years of professional experience, most of it in aviation. Starting from 2005, he mostly worked on Quality and Safety Management, especially in Risky Industries. His research interest is in Quality, Safety, Metrology, Non-Linear processes, and Calibration. His teaching experience is built in the Military Technological College in Muscat (Oman) and in GAL ANS in Abu Dhabi (UAE) where he taught aviation subjects (EASA Part 66 and ATSEP). At the moment, he is working as Faculty – Aviation Science at the Higher College of Technology (Khalifa Bin Zayeed Air College) in Al Ain (UAE).

He has attended many conferences and presented many papers. He has published three books in USA covering areas of Quality Management, Safety Management, and Quality/Safety Auditing.

Acronyms and Abbreviations

AFA	Anticipatory Failure Analysis
AFD	Anticipatory Failure Determination
AFP	Anticipatory Failure Prediction
AI	Artificial Intelligence
ALARP	As Low As Reasonably Practical
ATCo	Air Traffic Controller / Air Traffic Control officer
BBS	Behavior Based Safety
BE	Black Elephant
BJT	Bipolar Junction Transistor
BS	Behavioral Safety
BSe	Black Swan events
CEO	Chief Executing Officer
CF	Cynefin Framework
CPD	Combined Probability Distribution
CT	Computer Tomography
DKe	Dragon-King events
EE	Extraordinary Events
FAA	Federal Aviation Administration
FMEA	Failure Mode and Effect Analysis
FMEAA	Failure Mode and Effects Anticipation and Analysis
FMECA	Failure Mode and Effect Criticality Analysis
GMP	Good Manufacturing (Management) Practice
GRe	Gray Rhino events
GSe	Gray Swan events
ICAO	International Civil Aviation Organization
IG	Invisible Gorilla
ILS	Instrumental Landing System
IPD	Individual Probability Distributions
IQ	Intelligence Quotient
JPD	Joint Probability Distributions
KPI	Key Performance Indicators
LCL	Lower Control Limit
LSL	Lower Specification Limit
MIT	Massachusetts Institute of Technology
MLDMP	Machine Learning Decision-Making Process
MMS	Mineral Management Service
MTBF	Mean Time Between Faults
MTC	Military Technological College (Muscat, Oman)
NsNhNsM	No seen - No hear - No speak Monkeys
NTSB	National Transportation Safety Board
OHS	Occupational Health & Safety
OiS	Ostrich in the Sand

PANS	Procedures for Air Navigation Services
QMS	Quality Management System
RE	Resilience Engineering
RM	Resilience Management
RPN	Risk Priority Number
RST	Reverse Stress Testing
RUL	Remaining Useful Life
SMS	Safety Management System
TMA	Terminal Maneuvering Area
ToC	Theory of Chaos
TRIZ	Teoria Resheniya Izobretatelskikh Zadatch (Russian!)
TTL	Transistor-Transistor Logic
VaR	Value at Risk

Introductory Explanations Regarding the Things and Terms in This Book

Maybe you will wonder why I am speaking about all these things here, at the beginning of this book, but please understand that all these things will be used later in the explanations given for particular aspects of any of the "safety animals" mentioned in this book.

I consider this book as an unconventional book, so at the beginning let's clarify some of the things and terms used in this book...

This is a book regarding the metaphors used to describe some events which, in general, happen in our private and professional lives. These are metaphors for adverse events that correspond to some behavior of an animal or some impression associated with that animal. Everything started with the book *The Black Swan: The Impact of Highly Improbable* by Nassim Nicholas Taleb (NNT). This book uses a metaphor for Black Swan, having in mind that, for a long time, the humans believed that the swans could be only white, until the Dutch sailors saw the first black swans in Australia. NNT's book provoked the use of other metaphors, and most of them (named as "safety animals" by me) are covered in this book. Of course, many of these metaphors are used in our ordinary lives also, but this is the book where these metaphors are considered only from "the context of the things" in the Risky Industries.

If we speak about Safety Management System (SMS) as a systematic way to provide safety, then it can be described as the sum (compound) of Equipment, Humans, and Procedures. The Procedures are what connect the Equipment and the Humans and they actually build the management system.

When I speak about Humans, Procedures, and Equipment as constitutive parts of any management system in this book, I will use capital letters at the beginning of the words. For all other cases, I will use lowercase letters. This is important, especially when I speak about the equipment. Using the word "Equipment" with capital letter means that I am speaking about everything in the company that is not human: Machines, tools, instruments, cars, premises, furniture, etc. But using the word "equipment" with a lowercase letter means that I am speaking only about the systems used to produce, store, or maintain the product or to provide services. These could be: Trains, aircraft, cars, nuclear reactors, surgical tools, life-sustaining machines, manufacturing machines, etc.

The same things apply also to State (country) and state (phase of process); Procedure (System or Operating) and procedure (in general), etc. I tried very much to achieve this, but if I missed something, I do believe that "the context of the things" will clarify the use of the capital or small letters at the beginning of the words.

I will, very much, use in the book the term "adverse events" with the intention to provide explanation for events that can be characterized as incidents and accidents.

So, the "adverse events" is a common term for incidents and accidents. The accidents in the Risky Industry are defined as events where there is death of humans or catastrophic damage of the assets and/or the environment beyond possibility to be repaired. The incidents are defined as events where there are some injuries and/or a small number of deaths of the humans associated with particular damage of the assets and/or environment, but these are not so catastrophic. "Not so catastrophic" means that the damages of the assets and/or environment can be fixed and they can be used again.

When I use the words "Quality" and "Safety" with capital first letters, it means that I speak about the general areas of quality and safety represented the systematic way in the companies by Quality Management System (QMS) and Safety Management System (SMS). When I am speaking about Quality and Safety, I am thinking also on the Risk Management as an inevitable part of these two areas. Using these terms with lowercase letters (quality and safety) means that I speak about particular characteristic of something (product or service!) which is connected to quality or safety.

In general (and to simplify the future explanations), the "safety animals" are part of the problems created mostly by Humans.[4] But whatever we are considering in this book, it is highly dependent on the nature of the situation or that what I call "the context of the things".

In the book I will use the term "Stock Exchange" as a synonym for the areas where BSe are analyzed in NNT's books (for the areas of finance, investment, markets, politics, sociology, etc.).

In this book, when I would like to refer to Dr. Nassim Nicholas Taleb, I will use acronym NNT. This is done only for practical reasons and I do not have any intention to undervalue the personality of Dr. Taleb!

There is something which is a little bit tricky in NNT's book… He is speaking about the BSe which are discrete data (outcomes, events), if we consider the use of the language of statistics and probability. From another side, NNT speaks about Normal (Gaussian) distribution and other "Fat-tail" distributions which are actually produced by outcomes of continuous data. I can easily understand why this happens, but for the readers: This should not bother you! This happens because the quantity of data in Stock Exchange is huge and, in reality, it can be considered as continuous. My understanding is that he speaks more for the resemblance of the probability distributions than trying to be scientifically correct. And there is nothing wrong there if you know it. It is good just to mention that all events in our lives are discrete, so we can explain them as qualitative change of the states and processes in the industry which can be associated with some continuous change.

An additional thing I would like to emphasize here is the fact that in my life, I have not found a book about introduction to probability that is less than 400 pages and introduction to statistics that is less than 250 pages. So, please note that just the essential basics are mentioned in this book regarding probability and statistics. If you feel that you need more understanding about probability and statistics, you can attend

[4] They can be also caused by the equipment if the dynamics of change of states of the equipment enters the Chaos.

some of the video courses offered on the Internet or find some of the many books available for free download there.

In many places in this book, the Reader can notice that some things are repeating. This is not because I was confused when I wrote this book, but instead for two simple reasons…

The first one is hidden in the Latin adage: Repetitio ist matter studiorum (Repeating is mother of learning). That adage was part of my teaching strategy during my last five teaching years and I noticed that it worked very well with the students.

The second thing is that many of these "repetitions" are, actually, not "repetitions" due to the different "context of the things" where they are used. I hope the Reader can understand this when they read the book.

And the last thing: At the time of writing this book, I have spent around 25 years in aviation, so, please excuse the fact that I have used many examples from that area. However, I tried to present an attitude which apply to any Risky Industry, and I hope that the readers from any Risky Industry could find benefit in this book.

I thought all these explanations at the beginning of this book would be very important for the Reader who will continue with the rest of this book. I hope you will agree with my opinion later…

And finally, many of the complex graphic diagrams were produced using Desmos online graphic calculator (https://www.desmos.com/calculator). I really enjoyed working with it!

1 Philosophy of Science As Introduction

1.1 INTRODUCTION

Science is something that has contributed the most to human well-being. Whatever you think about science and the scientists, turn around and you can notice things that are products of the efforts of the scientists: Your laptop, mobile, car, food, clothes, home, etc. In everything, scientists have put their knowledge, experience, and strange and good ideas, and it resulted in all these things that we use today.

Not always was this fact recognized by the humans…

At the beginning of human journey throughout the centuries, the functioning of the nature was not so obvious to them. The mystic force in our brains was looking for the explanation of the things around us that we could not explain. Very often, the gathered knowledge showed benefit, so people tried to use it to improve their lives. But not always was the gathered knowledge describing the true nature of the world around us. Unfortunately, not all people were able to understand the things as they were, and many of them were prone to accept some other understandings about the nature and the world around them. They tried to pursue their ideas about functioning of the nature and the world to others.

I would not speak about the bad and the good things of the philosophy of general science in this chapter, but I would speak about the development paths of today's achievements and how the ideas of understanding the nature and the world were evolving in our history. Of course, everything in this chapter should be seen through the aspect of "safety animals".

1.2 HISTORY OF HUMAN SCIENTIFIC THINKING

The tremendous change in the human understanding of the nature and the world around us was achieved in the 15th century. The new geographic discoveries at the end of that century and at the beginning of the next century showed to the people that, even in this planet, there are different worlds and the things are not the same everywhere. This caused an emergence of the critical understanding of the nature and the world around us by the cleverest people and they started to do and publish their researches. These publications were a new dawn of new ideas. I would not like to take a side in the history, but I found very important for Renaissance in Europe, the fall of Emirate of Granada in Spain. Queen Isabel I of Castile and her husband King Ferdinand II of Aragon defeated the Muslim forces in Granada and Spain was Catholic again. In that time, the Islamic libraries in Granada and Alhambra were full with the saved works of old Greek philosophers. All these works became available to European scientists and all these erudite people recovered the

DOI: 10.1201/9781003230298-1

already lost knowledge of old Greek philosophers and spread them to the European public.

These Arabic translations of the Greek philosophers had a profound effect in Europe. The period of Renaissance started, and some other views of the nature and the world (never seen before) were presented to the people. The step forward from metaphysics to physics (or from religion to science) was done. In the next few centuries, the development of science offered the knowledge that brought benefit to most people, and this period is lasting even today.

The development of physics, which happened after Renaissance, contributed very much in building the machines that were reason for the first industrial revolution. This affected the social development also, but the pace of scientific and technological development was (unfortunately) faster than that of social development. The problem, which is present even today, was lack of development of the social and industrial skills of the humans compared to the development of the technology. Even today, when we are surrounded by all these engineering and technological marvels, our social development is still behind. The effect of engineering and technology made changes to our lives, but for plenty of us, it is just contribution to the Human Factors[1] (HF). And these HF will be also considered in this book from the aspect of the "safety animals".

In the time of Renaissance, the first changes of the mind and understandings were connected by the critics of the religious thinking.[2] The deterministic (causal) nature of the world was established later. It is not my intention to go deeply into the philosophical explanation of the Determinism because there are even different definitions depending on the context of the investigations or its use.

I will just explain Determinism as a way of understanding our world as governed by particular cause for every event and particular consequence for every cause. Of course, it is valid if the natural laws are causal. By using them, we can determine what will happen in the future by knowing the situation today and by using the data about situations from the past. It means that we can predict future events by knowing the deterministic laws of nature and by having data from the past and the present state of things. In mathematics this means: Having a formula and knowing the initial conditions, we can find the state of the system in the future.

Of course, that it is not so simple…

For example, the cause of an apple falling from the tree is Newton's Gravitational Law, but it will happen only if the weight of the apple (the force of gravity) becomes bigger than the force of keeping the apple by stalk connected to the branch of the apple tree. So, the cause of the apple's falling is connected not only to the Gravitational Law but also by the imbalance of forces in nature. This interconnection and additional "meddling" of the laws shapes our world even today. If we do not recognize these things on time, later, it can make our life harder. Anyway, life is not easy at all…

---

[1] Human Factors are outcomes of scientific approach to the aspects of human behaviors which is contributing to adverse events (incidents and accidents).

[2] Even religious view of the world was deterministic. In the Middle Age's religious dogma there were four types of causes: Material, Formal, Final and Efficient. Scientists just inherited this view and accepted it because their results of experiments showed correctness with the behavior of nature.

In the past (and even today), Determinism could not explain all events as people notice that there are things that happen randomly. By random events, we speak about events that happen by chance or, compared with Determinism, there is maybe a particular cause for this event to happen, but the outcome (consequence) cannot be predicted. The random event can be explained by throwing a dice. If the dice is well balanced, all six sides of a dice can show up equally and we cannot predict which one will happen with accuracy. Scientifically, I can say that, the probability of each of them to happen is equal and that is the reason why I cannot predict which one of them will show up after tossing.

In general, the random events are events of guessing and the cause of happening as these cannot say anything that will predict the future events by knowing the past and the present events. If I throw a coin and I get a "tail", throwing a coin one more time will produce a result I can only guess. Simply, the "tail" from the previous throw will not contribute to my understanding of the process, it will not help me predict whether I get a "head" or a "tail" in the next throw.

Approximately around 1960s, the world was deterministic and random, but in the second half of the last century, slowly, another "view of the world" was discovered: The Chaos!

At the beginning, the Chaos was discovered only in complex systems. Complex systems, in the general meaning of the word,[3] are systems that consist of many parts. The system functioning depends on the functioning of the parts, individually and in combination, with other parts, in the scope of the system. Actually, the system does not necessarily need to be complex (made by many parts). It could be also very simple, but it is enough if the interactions between the parts are complex. Most of the complex systems have interactions between the parts inside which are nonlinear, and as such, it is not easy to (scientifically) find the solutions of the equations that describe the system. In general, the Chaos can show up if the description of the functioning of the system can be described by three or more variables. If there are only two variables, the Chaos cannot happen.

The evident example of the complex system, where the Chaos can be registered, is the weather. Other examples are the turbulence in the air, the flowing of liquid through the pipes, the predator-prey model in nature, etc.

Usually, the complexity of the systems is connected by uncertainty of the prediction of how the system will behave: More nonlinearity – more uncertainty. And more uncertainty – more unpredictability.

The Chaos was noticed by the humans around 1960s as a purely mathematical concept and, as such, it was not accepted immediately by the physicists. Although there was some research dedicated to the nonlinear dynamics of the complex systems, during those times, the physicists mostly dealt with the theory of relativity and the quantum mechanics. But thanks to some of the "stubborn" scientists, the Chaos found its "place under the sun". At the beginning, it was not so revolutionary due to its abstract nature, but having in mind that it was dealing with nonlinear system's behavior, it developed very fast.

[3] There are also complex systems in regard to the equipment (machinery!), which may differ from this general meaning.

Later researches showed that the Chaos can be noticed even in simple systems with nonlinear structure. Of course, if the structure of the systems is more complex, the Chaos will bring more unpredictability. Today, there are Chaos researches in the fields of meteorology, physics, biology, chemistry, cardiology, economy, etc.

1.3 DETERMINISM, RANDOMNESS, AND CHAOS

The Determinism was the pillar of the science for many centuries in the past, and it is still present today, simply because it works very well. Whatever we think about the combination of nonlinearity in our world and Determinism, do not forget that nevertheless universe is ruled by the theories of Einstein's Relativity and the Randomness of the Quantum Mechanics; the space travels of many spacecraft (with or without humans) are made through application of Newton's Gravity Law, which is strongly deterministic.

Usually, scientists dedicate the particular phenomenon in the reality to a system, explaining that a system is a closed entity where this phenomenon showed up, moved, or worked. So, scientists investigate the phenomena in closed systems. Usually, there is a particular mathematical expression (equation) that is dedicated to each phenomenon as its explanation and the equation is used for prediction of what will happen with the system (phenomenon) in the future. This explanation of Determinism is like a recipe: You have a recipe for a cake and you cook the cake following the recipe. All further cooking of the cakes will produce the same cake, if you follow the recipe. So, if we are sure that you will follow the recipe, you can predict how the cake will be.

This example brings us to another characteristic of the Determinism: The events there are also repeatable: The cooking of the cake by use of the same recipe provides repeating of the same cake.

That is the reason that we say: Determinism is connected with predictability and repeatability. We know the initial condition of the movement (position, velocity, direction, environment, etc.), we know the equation (the law, the formula, etc.) describing the movement, so we can predict where the body will be in next few minutes, few hours, few days, etc. That is the actual mode of how our spacecraft navigates in our Solar system.

The problem with Determinism is that it applies only to linear systems or linear processes. The linear system (or linear process) can be described mathematically by a simple equation:

$$y = a \cdot x + b$$

where y is the output (outcome) of the system, x is input into the system (variable) and a and b are parameters that define the system. Graphically presented, this equation will result in a straight line in the graph and that is the reason for the name "linear system". If other equations for describing the system or the process are used, the graph is a curve (different than a straight line), then this equation describes a nonlinear system (process)

But applying the same linear equations for the molecules and atoms, it is impossible to predict their movement. In the world of gas molecules, although there is some kind of movement, we cannot predict this movement or their positions, simply because these are nonlinear systems and, as such, it cannot be described by linear equations. This is extremely valid for the particles in the nuclear physics: We cannot provide the exact position of electrons in the atom's orbits and their speed (moment) with the same accuracy[4] at the same time. This example is a beautiful example of transition (and connection) from one "world" to another. Actually, it is better to say: Transition from Macroscopy to Microscopy or from Determinism to Randomness.

The Randomness[5] is opposite to Determinism: We cannot predict the future outcome of throwing a coin or a dice. We can just dedicate the particular probability to each outcome and, by using this probability, we can guess the outcome. These are events that, actually, happen by chance.

The repeatability is also present here, but we cannot predict when the outcome will be repeated.

Either way, the tossing of a dice is connected with the laws of physics and, theoretically speaking, to calculate the outcome, I need to take care for extremely huge numbers of physical and mathematical laws. These laws are applicable on the initial conditions in the environment and they shall be applied at the moment when I toss a dice. I need to know these initial conditions and to use them as variables in the laws of physics, in real time for the mathematical calculations. Some of the important initial conditions for this case are: The force used to toss a dice, the angle of tossing, the mass and the elasticity of the dice, the distribution of the mass in the dice, the shape of the edges of the die, the angle of hitting the ground, the gravity law at the place of tossing the dice, the density of the air, the dice, and the ground, the pressure, the humidity, the temperature, etc.

In theory, the tossing of a dice is dynamical phenomenon and if I have all these data, (theoretically), I can calculate the outcome. But in reality, the system is so dynamic, so nonlinear, and so complex that I do not have the necessary equipment (sensors) to measure all data and I do not have the real-time computers that will execute all these necessary calculations. So, I can simply say that the outcome of the tossing a dice is a random event and I can only guess the outcome. Of course, the guessing is totally opposite of the prediction.

Why?

The prediction means I will have always the same outcomes if I know the laws and the initial conditions, and the guessing means that I will not have always the same event as I have guessed.

In general, the Determinism prevails in the macroscopic world and the Randomness prevails in the microscopic world.

The Chaos is not random, but although it is very much deterministic, we cannot predict the future in many cases due to the complexity and the nonlinearity of the change of the chaotic phenomenon. Although, there is a straight connection of the

[4] This is Heisenberg uncertainty principle applicable for particles in nuclear physics.

[5] In the literature you will find another term used as "Randomness" and this word is "stochastic". Roughly speaking, in the areas of complex systems and operations, the Random variables can produce stochastic process.

Chaos with one "safety animal", I would not speak here for the Chaos in details, simply because it is so complex that it can be a topic for another book.

Anyway, it is good to mention the "animal" here: It is the butterfly.

The "sensitive dependence on initial conditions" is the basic characteristic of the Chaos. We cannot always establish the accurate value of the initial conditions to any physical system in the nature. Always there is some small value that is missing, and all these small values contribute later to totally unpredictable behavior of the system under consideration. This is connected with the story which is the hallmark for the Chaos: The small movement of the wings of butterfly in Texas today will amplify later so much that it can produce a typhoon in China next month.

I am a strong believer of causality, but I am not a very strong believer in Determinism. I do believe that Determinism, Randomness, and Chaos are connected in a particular balance and their existence is actually a story of our world.

This is applicable (by my humble opinion) also to the nature: We can have a deterministic view of the nature, but complexity can arise, so we cannot use Determinism (linear equations) to explain the situation any more. This is strongly connected by the transition from Determinism to Chaos due to the sensitive dependence on the initial conditions that cannot be determined with the necessary accuracy (uncertainty is very big). In such a situation, the future state of the system cannot be predicted, and we can just guess what will happen based on a probability. This is something very much connected with Safety: The guessing is not what we try to achieve with implementation of Safety Management Systems in the Risky Industries.

There is another connection between the Determinism, the Randomness, and the Chaos. Nature is fundamentally based on evolution. The evolution is full of Determinism, but also Randomness is present in the same quantities. The small changes of living beings during their life cycles, by accumulation, can produce big changes in their development through centuries. This is something which I already mentioned above as "sensitive dependence on initial conditions," and it is part of Chaos. The big biological changes (mutations) can also interfere with the life of the beings in nature, and it can be considered as one of the factors of the adaptation of the living creatures to different environmental conditions. This is actually Darwin's evolution itself!

This mixture and balance of Determinism, Randomness, and Chaos in nature is one of the additional reasons why we cannot always predict events in the future.

1.4 "THE CONTEXT OF THE THINGS"

I do not think of myself as a scientist…

I was always dedicated to the science, but I missed my chance to become a scientist in the past. To be clear: I do not regret that because I never lost my touch with the science. On the contrary, I dedicated myself to applied science, which is called **engineering**. From time to time, I have noticed that the amount of science which I put into engineering is bigger than amount of science which is put by other engineers. So, I belong to nowhere. The scientists told me that I am too much "engineering" for science and the engineers told me that I am too "scientific" for an engineer. For the sake of truth, I feel myself in the middle (in the grey area in between) and the result is that nobody likes me and nobody listens to me.

Anyway, in this "gray area" between science and engineering, I found a beautiful field for research with an intention to improve the things in our real life. In my humble attitude, I prefer the "creativity of engineering" over the "science rigor".

My attitude (which is a mixture of Determinism, Randomness, and Chaos!) to the real life is something which I am calling "the context of the things". Simply, the science is for laboratory and the engineering is for industry (real life). But they do not cancel each other! There is a road that needs to be built for science and engineering to meet, and this road, even if it is built sometimes, it is not always followed.

The point with "the context of the things" can be explained using Figure 1.1. This is a simple example of the state of the water as its temperature changes.

At temperatures close to absolute zero (−273.16°C = 0°K), the water is solid. Actually, we can take a pot and put the water inside as a solid (hard) material. Make it sufficiently cold (no need to go too close to the absolute zero) and we can use this icy solid water as a hammer. Have in mind that this is a thought experiment;[6] in practice, we need good insulating gloves not to freeze our hands by using this "icy hammer".

By increasing the temperature, we will allow the water to reach temperatures higher than 273.16°K (0°C): The water will be transformed from solid into liquid. We use the water primarily as liquid, and our life is dependent on it. Going further, by increasing the temperature to temperatures higher than 373.16°K (100°C), the water will become a gas (vapor). Going further, let's say towards the temperature around 10,000°K, the thermal energy in the water (a gas) will produce enough energy for the atoms of hydrogen and oxygen to split and to leave atoms of the molecules of thc water (H_2O), so the water will not exist anymore as a matter. Actually, these are the

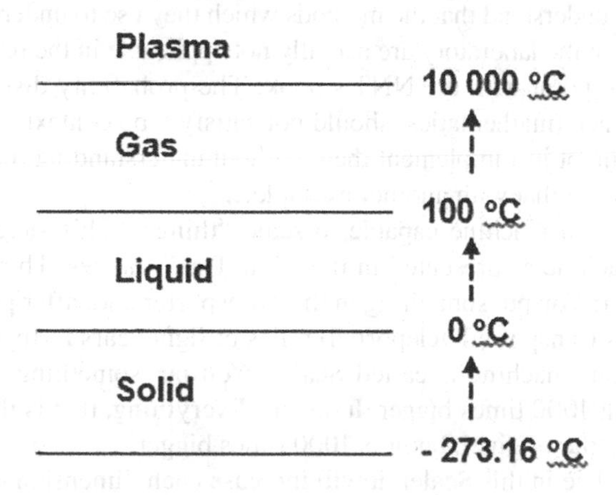

FIGURE 1.1 The state of water on different temperatures

[6] For the sake of truth, this thought experiment in reality is the subject of Clapeyron's equation and is pretty much complex. The needed simplification is presented here.

temperatures where the water stops to exist as water: There will be just a mixture of atoms, ions, electrons, protons, and neutrons called plasma.[7]

The main point here is that we have the same matter in four states and in all these states matter behaves abiding to the different physical laws.

a. In the state of ice, the water will behave as a solid material (subject of pretty much deterministic laws);
b. In the two other states, it will behave as deterministic (as a whole) or random (as atoms inside the gas and liquid); and
c. In the fourth state (as plasma), it will behave totally chaotic.

Which laws will be used to describe the water on different temperatures also highly depends on the situation (rest or move). So, if we like to understand water in total, we should know each type of state of the matter (or its behavior) and the associated laws to these states. And this is something which I call "the context of the things". Depending on "the context of the water", you choose different methods (physical laws) to explain its behavior.

As can be seen in Figure 1.1, it is described as the same matter (water!) that is put in the same environment, but the temperature of the matter ("the context of the things"!) will determine which laws will be used to depict the situation with the water!

In general, the criteria in the laboratory do not fit the criteria in the real life simply because there is a different context between these two places. The laboratory is a controlled environment where we can control the interactions, but the real life is filled with uncontrolled interactions between the users, the environment, and the products (or services). This is a gap ("the context of the things") that is not always recognized by the scientists and the engineers. You may think that this is a stupid conclusion, but I have met plenty of engineers and scientists in my life and I am shocked with the number of them who did not understand that the methods which they use to understand or to control the situations in the laboratory are actually not applicable in the reality.

And this is the point with the NNT's book: The probability distributions determined in the science (mathematics) should not satisfy "the context of the things" of the reality. We cannot just implement them without understanding the situation!

Let me emphasize that with another example…

Assume there is a machine capable to scale "things". This situation is similar to the kind of machine as presented in the "Star Trek" movies. There is a machine called Transporter: You put something in the Transporter and, after pressing the button, the "thing" is transported (teleported) miles or light years farther away.

In this case, our machine is called Scaler: You put something in the machine and it will make it 1000 times bigger. It means: Everything, that is the output of the machine, looks as the original, but it is 1000 times bigger.

So, if we put a bee in this Scaler, it will increase each dimension of our bee 1000 times. Imagine such a machine: The bee scaled this way will provide 1000 times

[7] Plasma is known as fourth state of the matter. It can be defined as a bunch of ions and electrons in a cloud that behave in a collective manner making a cloud of plasma neutral. It means that, in the state of plasma, the water is not existing anymore as water.

more honey than the normal bee. It will solve the problem with the starvation in the world because, if we put chickens in there, we will produce 1000 times more chicken meat. Or with bread, eggs, milk, etc.

But the simple question is: Will the "scaled bee" be capable of flying?

The answer is simple: Not at all!

Actually, the bee will not be capable of even staying on its legs: They will break. The tissue inside the normal bee is in accordance with the size of the bee. If we simple scale it, it will simply not work. Scaling the bee has no sense because it will produce completely different "context of things" and the "building material" of the bee's legs will not survive 1000 times bigger weight.

This simple example applies to the industry and by taking into consideration "the context of the things", it will have the following consequences: You cannot implement the same structure for the safety (quality, environmental, security, etc.) management system for a company with 50 employees and for a company with 1000 employees. The overall context of their operations, basically and generally, is different!

The point is that "the context of the things" put us in a position to take care of the applied mathematics in the area of the real life instead of pure mathematics. The pure mathematics[8] applies only to the laboratory! So, in our reality, by taking care of "the context of the things", we should choose: Pure mathematics (science) or applied mathematics (real life or engineering)?

Let's support this with one simple "scientific" example:

To explain a term "function" mathematically, I will state that the function $f(x)$ is a rule for mapping the elements from one set into the elements of another set. Assuming that it is mapping from the set of real numbers (R) into the set of real numbers (R), I can write (mathematically!) the quadratic function as:

$$f : R \rightarrow R => f(x) = x^2$$

This is real mathematics, but in engineering (real life!), there is no engineer who will use this notation. Everyone will say that the function is x^2. For those who are not good in theoretical mathematics, please understand that x^2 is number, not a function. Or speaking more "picturesque", the $f(x)=x^2$ is parabolic line and x^2 is only a point for any particular value of x on that parabolic line.

1.5 REASONS FOR UNPREDICTABILITY

When I speak about the prediction of the adverse events, I would like to know which type of event and when it will happen. So, knowing WHAT will happen is not enough. I need to know also WHEN it will happen. Only by knowing WHAT and WHEN, I could prepare myself.

[8] For the purpose of this book, when I say "pure mathematics" this is mathematics as science by itself (solving mathematical problems, proving theorems, etc.). When I say "applied mathematics" this is mathematics which is used with the knowledge in any other area to solve or explain the situation, status or problems in this area. Probability is one of the oldest representatives of the applied mathematics.

The main reasons for the unpredictability are that, due to different factors, there is always uncertainty of how and when the system will move from one state to another. The reason for this uncertainty can be anything that is associated with the uncertainty of some of the parameters or variables in the system or in the environment around. This uncertainty could be behind the possible faults in the systems or the failures of operations. In other words: Whenever uncertainty is present, there is a risk that something will go different than expected!

The uncertainty, same as the Randomness, affects the predictability of the systems, so let's look in more detail at Randomness and Uncertainty.

1.5.1 Randomness

NNT explains the BSe[9] from the point of Randomness. The very important point with this explanation is a question: Can the Randomness produce something which was never seen and never experienced or it is just connected to the well-known set of events?

In our safety case, these set of events could be states of the system (equipment, process, operation, etc.) that show up randomly, so we cannot predict which state will be the next one. Obviously, the Randomness itself does not guarantee some novelty (or some type of BSe that were not seen before). Mostly, in science and in life, Randomness is used on a set of well-known elements[10] that pop-up randomly.

However, let's see, in more detail, what goes on with the Randomness in different areas...

The Randomness is a term used mostly in the probability and in the statistics. When I searched on Internet for the term "Randomness", I found approximately 14 million hits. In general, the term Randomness can be clarified as one of the reasons for unpredictability. We can see later that there are also other reasons for the unpredictability.

As mentioned before, the predictability is a part of Determinism. It is connected to some existing logical structure described by an equation (or by set of equations) that can be used to describe the event (state of the system) mathematically. From other side, the lack of such equations that describe some process (event, system), could be used as a definition of Randomness. Having this in mind, we can say that the lack of Determinism in the events is a sign of Randomness. Mostly in science, the research of the system looks for a particular level of Determinism, and when any type of Determinism cannot be noticed, Randomness is declared.

Graphically (scientifically), the difference between a random and deterministic process can be presented as in Figure 1.2.

On the left side on the Figure 1.2 are presented the states (each dot one state) of a random (stochastic) process (x is input and y is output). On the right side are presented states of a deterministic process. Intuitively, the states on the left side cannot be connected with any rule (equation) and states on the right side

[9] In some of the literature available regarding the BSe you can find explanation that the existence or otherwise of Black Swans is not an event, but a state of affairs. I will use name of Black Swans as events in this book.

[10] More about that in Section 2.3.1 (It Is Unpredictable...) where probability will be explained.

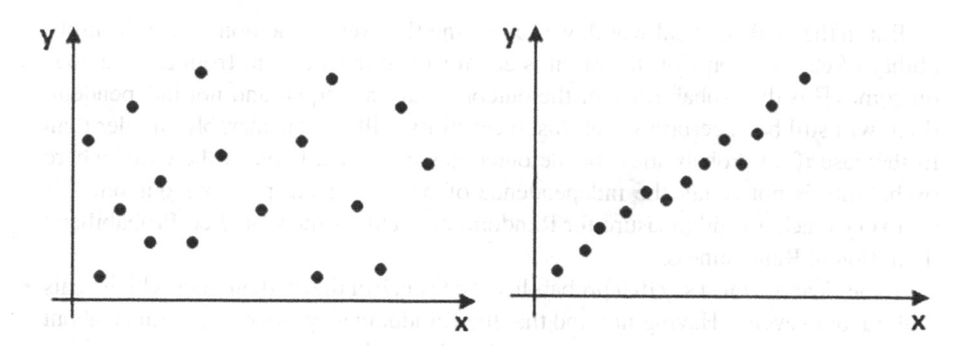

FIGURE 1.2 The difference between random (left) and deterministic (right) processes

(nevertheless they are not strictly aligned) can be (approximately) depicted by the equation:

$$y = kx$$

where y is the output of the process, x is input (variable) in the process, and k is parameter. If this equation is applied in electronics, especially describing the ratio between the input and output of the amplifier, k will be the gain of the amplifier.

The equation above describes a line, but looking of the dots on the right side of the Figure 1.2, we can notice that they are not strictly aligned into a line. There is simple explanation for this non-alignment: There is uncertainty in measuring and determining the state (presented as a dot) on the system, so this uncertainty about the position of the dots (states) is the reason for the non-alignment. This is also an explanation for why the Determinism is also a subject of particular uncertainty. In the perfect world, the uncertainty would not exist. But in the real world, we must live with it.

It should be emphasized here that, not always, the Randomness in the data can be recognized. And this is very much important for the Risky Industries. There, the risk calculations are based on samples of data, which cannot be always enough big to determine if it is a random or a deterministic process. There are many statistical tests for the purpose of determining the Randomness in the statistics, but it is not the intention of this book to go through them.

So, as I have said before, in the real world, some event[11] is random if there is no recognition of the particular rule of how the event occurs. In the industry, we use Randomness when we cannot predict the state of the process or the equipment when there is a fault. A simple example of the Randomness in digital electronics is SR flip-flop. It has two inputs (S and R) and two outputs (Q and Q'). Every engineer of electronics is aware that, in the cases when the inputs are S = 1 and R = 1, the outputs are random and we do not know what the output of the SR flip-flop will be.

[11] For the purpose of this book, the term "event" means "adverse event" and the term "outcomes" means "different types of adverse events". In the most of the literature, dealing with random processes and probability, the "event" is actually outcome from experiment. So, tossing a dice is an "experiment" and outcomes "head" and "tail" are "events".

But in the mathematical world, we can define the event as random when the probability of each outcome of this event is equal to and independent from the previous outcomes.[12] If the probabilities of the outcomes are not equal and not independent, there will still be uncertainty, but this uncertainty will be considerably smaller than in the case if all probabilities for the outcomes are same. Even in the cases where probability is not equal, the independence of outcome from previous outcomes is still very much a valid measure for Randomness. This is the so-called Probabilistic definition of Randomness.

To be clearer, I must say that probability is a branch of the mathematics which deals with random events. Having in mind that the Randomness produces ignorance about the future states of the system, it means that the random system (process) is unpredictable. That is the reason that we are using probability to describe Randomness by assigning a number (percentage) as a quantity of chance that a particular outcome (event) will happen.

As example of a random event, in accordance with the probability theory, could be the tossing a coin or tossing a die. In the case of a coin, the probability of getting the outcome "head" is same as the probability of getting the outcome "tail" (1/2 = 50%). In the case of tossing die, the probability for getting each number from 1 to 6 as outcome is same and it is equal to 1/6 (approximately 0.1667 = 16.67%). So, tossing a coin and tossing a die are random events in regards to the outcomes of each tossing because the probability is the same for each outcome. Most important, regarding the Randomness, is the fact that any next tossing is not connected to the previous tossing in both cases (coin or dice). It means if we toss a die 6 times and all six times, we get number 5, for the next (seventh) toss the probability to get number 5 is the same: $1/6 \approx 16.67\%$.

In the industry, there is a white noise in telecommunications which is totally random regarding the frequencies which it can show up. It means that the probability distribution regarding the frequency of the noise is uniform[13] (equal probability for each frequency).

In the 1960s, another definition of Randomness was published, and it is based on the information theory. The definition was based on the ideas of Solomonoff, Chaitin, and Kolmogorov, but its application is mostly for theoretical purposes.

The information theory says that when we transmit any kind of message, each message is part of the known set of messages. Let's say, when we pass the message to our friends, we speak the same language, so our friends can understand what we speak about. In other words, we use the language (letters and numbers) to "code" the information. Different languages – different "codes"!

[12] There is a big difference of accepting this definition of Randomness between the experts. NNT does not agree with this definition. He states that if even the probabilities are different, it can be random process. But, in the same chapter where this is stated, he is using example of Russian roulette which is pure random process with same probability for each number. Some of the scholars limit this definition to the finite number of events, which is understandable because, if there is infinite number of events (by mathematical definition!) probability for each event is zero. Anyway, I like this definition because it is connected by truly random events (tossing a coin or tossing a die).

[13] Do not mix this probability function with the probability function of the amplitudes of noise in electronics. The amplitudes of the noise abide to Normal (Gaussian) distribution, but the frequency probability function is uniform.

The letters and the numbers of our language are our "coding material" used by us to communicate messages. In other words, for the communication to be reasonable, there is a need the "transmitter" and the "receiver" to "speak" the same language (code).

Imagine two binary digital messages[14]:

010101010101010101010101 and
011000101110110011010010

By intuition, we can assume that the first message is deterministic and the second message is random. If I try to send the first message, I can send it in original (all 24 digits) or simply as "Twelve times 01" which is a shorter ("compressed") version. As can be seen, everyone will understand the shorter version of the first message, simply because I send with the message, the rule how this message was built.

If I try to send the second message, I must send it in original. Simply put, I cannot make it shorter (I cannot "compress" it) because I do not know the rule. If I try to make it shorter without the rule, the sent message will lose some of the digits and will lose information. In other words: The "receiver" will not be able to re-create the message when it is received.

According to these examples, the Randomness could be (roughly) defined as non-capability to shorten ("compress") the "messages" without losing the information. This definition is known as "algorithmic definition for Randomness".

In reality, there are processes which are random and there are computers which are used to simulate these processes. The Randomness is used very much by the Monte Carlo method for simulations in science and in industry. There, the experiments are simulated by random inputs and the results (outputs) are analyzed. But there is a problem because computers use programs for their operations and we cannot make a program that will make computer to do something by chance. A computer follows written instructions from the program blindly and is, therefore, completely deterministic.

The point about this is that we, actually, simulate the random inputs for the Monte Carlo simulations inside the computers. There are various computer applications that run these simulations, and thus they save time. The programs differ by the way of selection of random inputs for the events, because they are generated by, so called, "pseudo-random generators". The name "pseudo-random generators" comes from the fact that the random variables in these applications use a periodic mathematical equation to generate those values. After one cycle, the pseudo-generator starts generating the same numbers from the previous cycle. Periodic means repeatable, which means, they are actually not random, but deterministic. In Table 1.1, there are values of several pseudo-random generators and the periods of their cycles used as programs by the computers.

Looking the Table 1.1, you will assume that MT19937 is the best pseudo-random generator, simply because it provides the longest period of repetition of the numbers.

[14] This is explanation similar to one used by G.J. Chaitin in his paper "Randomness and Mathematical Proof", published in Scientific American 232, No.5 (May, 1975), pp. 47–52!

TABLE 1.1
Different Types of Pseudo-Random Generators Used for Monte Carlo Method

Type of Pseudo-Random Generator	Relative Speed	Period
RAN0	1,0	~ 2^{31}
RAN1	1,3	~ 2^{36}
RAN2	2,0	~ 2^{62}
RANQD2	0,25	~ 2^{30}
MT19937	0,8	~ 2^{19937}
TAUS	0,6	~ 2^{88}
RANLXD2	8,0	~ 2^{400}

This is true, but for scientific purposes (where there is need to repeat the experiments), the pseudo-generators with shorter period are also very useful. The Randomness is, very much, used by the computers, especially for video games and for encryption of data. There, the longer period is better.

Having in mind that Randomness is necessary in the science and also in the industry for research purposes, there are websites on the Internet that can provide services (paid or free) for generation of random numbers for different purposes. They usually use natural random phenomena as radioactive decay (https://www.fourmilab.ch/hotbits/), ambient, atmospheric, or white noise generated by electronic radio receivers (www.random.org), etc.

1.5.2 Uncertainty

The uncertainty is another cause of the unpredictability connected with the events. The uncertainty has the same intrinsic influence in our real life as the Randomness, but there is a difference. Although the probability is also used to describe the uncertainty, the uncertainty is not connected by chance or by guessing.

I will use the process of measurements to make a point about the uncertainty.

Trying to explain of the students in the Military Technological College (MTC) the basics of measurements, I gave a ruler and paper ribbon to each. I asked them to measure the length of the ribbon. They had to do that and the results should be written on the small piece of paper, which was given to each of them. Later, on the Excel sheet (prepared in advance by me), all 28 results were written and calculations and graph were produced.

There, what students could notice was that, although each of them received the same ruler, used the same method and the same ribbon, the results were different. On my question, why the results differ, they could not find an answer. My explanation at that time was the same as the one which I will provide here: The Metrology, as science of measurements, is clear that the measurement is an operation (activity, process) that cannot produce accurate number! There are plenty of reasons, and all of them are connected with some type of variability of the assets used during measurements: Instrument, measurand,[15] environment, measurement method, operator, calibration, time, temperature, etc. All these variabilities will contribute to the particular uncertainty of the result.

So, in the science of measurement, always the results are statistically calculated where (roughly speaking), the mean (**μ**, the arithmetic mean value) of these plenty of measurements is a measure of the accuracy of the quantity measured and the extended standard deviation (usually 3σ) is measurement for the precision. Having this in mind, the standard deviation could be defined as measure for uncertainty.

The result R of n measurements ($M_{i=1...n}$) of any measurand will be presented by the equation:

$$R = \mu \pm 3\sigma$$

where μ and σ would be calculated by using these equations:

$$\mu = \frac{\sum_{i=1}^{n} M_i}{n}; \quad \sigma = \sqrt{\frac{\sum_{i=1}^{n} (\mu - M_i)^2}{n-1}}$$

Let's explain this in more detail...

Doing plenty of measurements, we produce set of different values which will have some probability distribution, simply because we use the same process. In Figure 1.3 is presented a "virtual" probability distribution of many measurements expressed as Normal (Gaussian) distribution.

On y-axis is presented the frequency of the value of the measurement, and on the x-axis is presented the difference between the mean (μ) and the actual measurements ($M_{i=1...n}$). The shape of the distribution says that most of the measurement results will be close to mean (μ) and that a very small number of the measurements will differ significantly ($\mu - M_i$) from the mean (be far away from it). In other words, if there are many measurements with big difference ($\mu - M_i$) from the value of mean, it will result in low precision of this measurement system.

In the scope of the time, it will happen not very often under assumption that:

a. The measurement system is in "good shape" and properly calibrated;
b. The measurements are done in the laboratory under controlled environment (no changes in the temperature, humidity, pressure in time, or no changes in any other parameter which can affect the accuracy of measurement);
c. The method used for measurements is proven to be adequate;
d. The laboratory staff is trained and competent in using the measurement system and measurement method;

---

[15] Measurand is characteristics of the unit (thing, characteristic, etc.) which is subject of measurement.

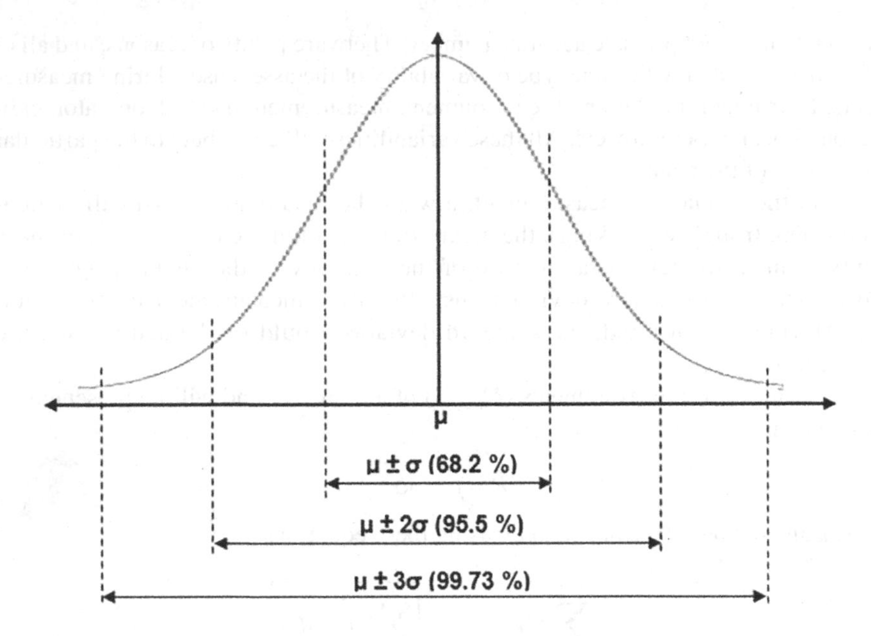

FIGURE 1.3 Probabilities of having the true value of measurement in different ranges of standard deviation (σ)

You will maybe ask: Where is here the probability?

The probability is hidden in the words "extended standard deviation" or, relying on the first equation, inside the expression $\pm 3\sigma$. As it is presented in Figure 1.3, it can be seen that there are three areas of consideration around the value of the mean (μ):

1. The probability that the true value of the measurement (out of the set of infinite repetitions of the measurements) will be in the area marked by $\mu \pm \sigma$ will be 68.2%;
2. The probability that the true value of the measurement (out of the set of infinite repetitions of the measurements) will be in the area marked by $\mu \pm 2\sigma$ will be 95.5%;
3. The probability that the true value of the measurement (out of the set of infinite repetitions of the measurements) will be in the area marked by $\mu \pm 3\sigma$ (known es extended standard deviation) will be 99.73%.

As it can be seen, if we express the measurement results as in the first equation in this paragraph (with $\mu \pm 3\sigma$), uncertainty will be only 0.27%. It is small, but it still exists. The main point is that for each value of the standard deviation σ (left and right from μ), there is a particular probability that our true value of the measurement will be. This probability is a measure of uncertainty regarding the true value also.

I showed above, the example which is using the Normal (Gaussian) distribution, just for the sake of simplicity, but in the realm of measurements, things are different…

If the number of measurements is approaching infinity, this probability distribution will be Normal (Gaussian), but in the most real cases when we limit the measurements to 10–15 tries, the probability distribution which will be more appropriate is, this is the so-called, **Student's t-distribution**.

This is a probability distribution that, very much, resembles the Normal (Gaussian) distribution, but the Metrology has proved in the past, that it corresponds better to the small number of the measurement results than the Normal (Gaussian) distribution.

This distribution was developed by William Sealy Gosset (1876–1937, English statistician, chemist and brewer) under the name **t-distribution**. Gosset published his papers under the pseudonym Student and that is the reason for the name of distribution today (Student t-distribution). Although it is very much used in Metrology, there are some articles on the Internet which state that this distribution (with 3 or 4 degrees of freedom) and Laplace distribution can be used in the Stock Exchange because they fit better than the Normal (Gaussian) distribution. If used there, the happenings of the BSe will be decreased.

In general, the Metrology uses statistics and probability to express our ignorance about the results of the measurements (uncertainty) due to the imperfection of the method, the instruments, the operator, or whatever else which could affect the measurement result.

The uncertainty can be also associated to the humans who must make decision, but some piece of information is missing. In such a case, there is not enough information to make the right decision. The decision can be clearly determined, but the process of making the decision is "shadowed" by the uncertainty of the data available. In this context, the definition of the uncertainty is very much close to the information theory.

In general, the theory of information says that we know all information which should be transmitted, but the order of information during transmission (what and when will be transmitted!) is unknown (uncertain). So, when we receive some kind of data, we receive the transmitted information hidden into data and our uncertainty is gone. We may say that: The receiving of data always decreases the uncertainty about the sent information!

In our daily lives, we use different synonyms for uncertainty:

a. Variability is used in the dynamic systems where there is not enough information about the changes of the states of the system. From "the context of the things" in Safety Management, this is very much important for the Risky Industries;
b. Vagueness (fuzziness) is used in ordinary lives when the words used are uncertain by its language definition. Let's say, the word "mature" is connected with psychological development of the person and as such it is associated to different age for each person. One can be mature at 16 years and one at 23 years;
c. Imperfection is used in the life and in the industry where we try to provide quantitative data about something which is descriptive by nature. In such a case, we are not sure about the characteristics or situation. For example, we use the description "good product", but what does "good product" mean in this situation?

d. Ambiguity is used when you cannot clearly understand the meaning of the words put into the sentence. It could happen very much when someone speaks a foreign language, but his knowledge of the language is not so good;
e. Etc.

In general, the science says that the uncertainty can be present in one of two ways. The first one is where we do not know the facts and we could not provide reasonable probability based on the available data. This usually happens when there is not enough data or the data are random. The second one is when there is no known method to determine the rule that could help to calculate the numerical value for probability. This usually happens when there are new phenomena discovered in the science and this happens, especially, in the quantum mechanics.

1.5.3 Randomness Vs Uncertainty

As I said previously: uncertainty is not Randomness!

Uncertainty, in the case of measurements, is described by standard deviation (σ) and is used as a measure for the precision of our measurements. Bigger standard deviation – bigger uncertainty.

But is it true?

Can we put inequality between the Randomness and the uncertainty…?

Honestly speaking: It is hard to say…

When I went on the Internet and tried to find answer there, the search hits on Google Chrome numbered around 7 million results. There were posts that put equality between these two terms and there were posts which state that they are different. For this book I need to take a side, and the text below expresses my attitude (which I will use it in the book) to both of them.

The Randomness is a characteristic of the processes and, as such, it is very much investigated by statistics, mathematics, science, and industry. We have random processes that are subject to random variables inside these processes. This use of the Randomness is with intention to make a difference with the Determinism, which is main subject of the science.

Assume that we have a shooting competition with a gun. If we follow the shots of one of the competitors, we can notice that most of them will be in the center point of the target, but there will be also the shots which will not be in the center of the target. I can say that the spreading of the shots looks very random and, as such, they will contain particular uncertainty.

If we have enough data (shots), we can calculate the uncertainty of the shots of this competitor. Probabilistically, it will be (maybe, but not surprising) the Normal (Gaussian) distribution with very high peek the center of the target. If the peak is sharp, the person is a good competitor; if the peak is not sharp, the person is not so good of a competitor (Figure 1.4). This is not a Randomness at all because the aim of the competitor is to hit the center of the target. So, having a peak or Normal (Gaussian) distribution is expected in such a case, but there is only uncertainty, not Randomness at all.

Surprised…?

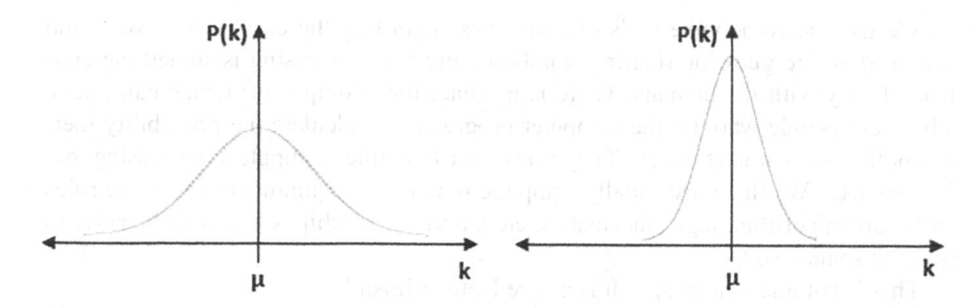

FIGURE 1.4 Results of bad competitor (left, lower kurtosis!) and results of good competitor (right, bigger kurtosis!)

No need to be surprised…

Imagine: You play in casino one of the gambling games based on tossing two dice.[16] It is your next hand and you need two 6s to win a lot of money. Of course, that you will strive and put all your efforts to toss the dices with intention to obtain two 6, but will it work? In the case of a gun competition, the training will bring you better results, but the training does not make the difference in the gambling (tossing a die). It does not make a difference because the outcomes do not depend of our skills or our efforts: The random process (tossing a dice) cannot be interfered with our wishes or our training or our skills or our money!

Using the explanations from previous paragraph, the kurtosis[17] in the Figure 1.4 can be used only to describe the uncertainty of the process through standard deviation (σ): The bigger standard deviation (wider curve!) – the lower kurtosis (and vice-versa). The kurtosis cannot be used to describe the Randomness because the Randomness cannot be measured.

OK, for the sake of truth, it is not so strong a statement because in the area of the Stock Exchange, there is a method known as Approximate Entropy[18] which looks for patterns in sets of data. These patterns can be created by the human intention to earn money by the Stock Exchange trading, but the patterns cannot be created with particular certainty in the industry.

The patterns in the Stock Exchange are generated by a rule: Buy when it is cheap and sell when it is expensive. Of course, these patterns will repeat themselves from time to time. By definition, Randomness could be explained by lack of patterns, so you can understand the nature of this method: It is clear that it cannot be used with required uncertainty in the industry for practical purposes.

Having in mind all this explanation, we can understand why the random processes (data) can be described by uniform probability function. The uniform probability function has no particular patterns that repeat. Any pattern there could happen with the same probability.

---

[16] Sorry, but I am totally unfamiliar with the casino matters and I have no clue what is the name of this game. I have just seen it on the movies…

[17] In statistics, the term "kurtosis" is used to describe the "sharpness" (or "tailness") of the probabilistic distributions: Sharper peek (smaller tails) means bigger kurtosis

[18] Approximate Entropy is a method which is used also in psychology and some areas of finance.

We use the Randomness also in our lives regarding the cases when we would like to describe guessing (betting, gambling, etc.). The guessing is something connected only with the humans. There is no machine (Equipment) which can guess. There are people who use the computer programs to calculate the probability their football team win next match. They think that machine (computer) is guessing, but it is not true. We (humans) usually equip the machines (Equipment) with some rules (software, algorithm, logic, mechanics, etc.), and the machines follow these rules to provide some results.

This is not guessing at all – It is a pure Determinism!

As said in previous paragraph regarding the Randomness, there are many computer applications[19] that can produce (periodically) random numbers, but these are actually pseudo-random numbers. These applications (based on periodicity) produce deterministic numbers, but their number (before they start to repeat themselves) is so huge that in the short time, we can assume that they are random. Obviously, in such a condition, the uncertainty how they will show up on the screen is very big.

This is not a case with pure uncertainty.

I will repeat again: Speaking about the uncertainty, especially in terms of measurement, we speak about our ignorance regarding the result. The reason for that ignorance is not the Randomness of the measurement as a process. Simply, we do not know what is the accurate result because we do not have enough data and we can express this uncertainty through the mean (measurement for accuracy) and the standard deviation (measurement for precision).

The difference between the Randomness and the uncertainty is very much present in the quantum mechanics. There is the Heisenberg principle of uncertainty, which states that we cannot measure the location of the particle and its momentum simultaneously. Striving to measure the location, we are losing knowledge about the momentum and, vice versa, looking for the momentum, we are losing the knowledge about the location of the particle. I read somewhere that there is no Randomness in the quantum mechanics, but uncertainty is very much present!

However, even in quantum mechanics, the situation is not clear regarding the Randomness and the uncertainty. Writing this book, I found an article in Quanta Magazine which stated that one of the first applications of the quantum computers will be generating random numbers. So, in this article, the uncertainty is synonym for Randomness. Anyway, the radioactivity is very much full with uncertainty, but it is created by the Randomness, not by the Determinism.

There is another thing which will help us to make a difference between the Randomness and the uncertainty: The Theory of Chaos.[20]

The chaotic processes are strongly deterministic processes with huge uncertainty due to exponentially developed "sensitive dependence on initial conditions". There,

---

[19] The computer (and other!) applications for producing pseudorandom numbers are known as "pseudorandom generators". The simples one is RAN0 with a period of 2^{31} numbers and the most complex is MT19937 with the period of 2^{19937}.

[20] The Theory of Chaos will be subject of attention later in the book!

speaking about two moving micro-particles which are very close to each other, the uncertainty of their movement can be expressed by the equation:

$$\delta(t) = \delta(0) \cdot e^{\lambda t}$$

where $\delta(t)$ is distance between two close particles after particular time t; $\delta(0)$ is initial condition (initial distance between these two particles at the beginning, which is of the range of 10^{-4}); e is Euler's number (2.17828…), and λ is the so-called Lyapunov exponent.

As can be seen in the equation, these two particles (very close to each other at the beginning!) are exponentially distancing themselves. It means that uncertainty of their position is exponentially increasing.

So, the difference between the Randomness and the uncertainty exists. The Randomness always has a particular amount of uncertainty inside, but the uncertainty itself does not necessarily mean Randomness. This is the reason that categorizing the events as random or as uncertain, does not make sense.

In general, for the purposes of the Safety Management System (as "compound" of Equipment, Humans, and Procedures), I will use the Randomness when I would like to describe influence of Humans on the safety events. The reason is: The Humans are very often urged to make decision without enough information (especially managers). I will use the uncertainty as a characteristic of the industrial or other processes, when there is the use of Equipment.

The procedures could be also uncertain (ambiguous) if they are not prepared well, but they cannot be random. However, if something is not clearly determined by the procedure, the Humans (who are using the Procedures) must guess (improvise) what to do, which means Randomness. So, in the Risky Industries, which are the subjects for the implementation of the Safety Management System, the Randomness and the uncertainty could be present in same or in different areas.

I would like to warn the Readers about usage of the Randomness and uncertainty in the literature and in the real life: Be careful, very often these two terms are used as synonyms. I hope that the explanations in these few paragraphs will provide enough knowledge to make a difference, at least for the purpose of this book.

1.6 RANDOMNESS, UNCERTAINTY, AND PREDICTABILITY

Randomness and the uncertainty are factors that affect the predictability.

To be honest: There is no limit connected with the predictability. We can try to predict almost anything. We can try to predict even the outcomes of the random processes (dice, coin tossing, etc.) and things with high uncertainty, but the point is in the question: How valuable will be this prediction?

We can define the prediction as a statement regarding the outcome of something before it happens and in the presence of uncertainty (due to lack of information or any other reason). This definition deals with the prediction as a prophecy.

Very often, the prediction is based more on our wishes or our expectations than on something tangible. One of the characteristics of most of the humans is that, in the situation where we do not have enough information (data), we include our personal bias into our prediction regarding the outcome of something. To be more specific: Sometimes the bias is triggered by our wishes (optimistic) and sometimes by our fears (pessimistic).

Throughout human history, there have been many wrong predictions that resulted in huge damage to assets or many deaths of humans, and a well-known one is the Fukushima disaster. There was a prediction that a possible tsunami triggered by a possible earthquake will not be bigger than 5.7 m, but no one knows what was the method used for this calculation. The defense wall was built taking into account this calculation for the tsunami, but the tsunami which hit the plant that day, had a height of 14–15 m.

The humans use predictions for the future and for the past, and, in both cases, there is huge uncertainty. The simple example about prediction for the past events is this: If I had this information two years ago, I would not have bought that car.

I am not sure that it will work…

How would you know now what you have been thinking two years ago and what you would do in that time if this information was available to you?

There is something that most of the people are not aware of: The prediction, without statement of probability regarding its accuracy, is not a prediction. It is just a statement whose value cannot be determined, but this is mostly valid for the science and the engineering fields. For real life, things are simpler: There is no need for probability. As an example: Stating that tomorrow it will be a rainy day is too general, but if I state that chances for tomorrow to be a rainy day are 80%, it means that the probability that tomorrow will be a rainy day are four times bigger than the probability that tomorrow will not be a rainy day. In such a case, it is wise to bring an umbrella with you. But the point is that, nevertheless, the prediction is 4 to 1 (80%) in a favor of a rainy day: it can be that the sun shines tomorrow or that there are clouds but no rain.

Having in mind the previous paragraphs where the Randomness and the uncertainty were explained, I can say that the Randomness is not a characteristic of the prediction but that prediction is part of the Randomness. The prediction together with the probability, actually, can be associated with Randomness if the additional clarification of how big the Randomness is resolved.

Regarding the uncertainty, things are different. Uncertainty is the characteristic of the prediction, whatever the nature of the prediction is. The prediction is used for the things where there is lack of information or the uncertainty of available information is too high. Probability is used to quantify (express by number) the uncertainty which is associated to the prediction. This means that the use of the prediction is mostly, as a decision-making process, based on particular probability.

2 The Black Swan Events

2.1 INTRODUCTION

The book *The Black Swan: The Impact of Highly Improbable* from Nassim Nicholas Taleb[1] (PhD) was published in April 2007 by The Random House Publishing Group (USA), and it garnered massive interest, mostly among the traders and employees in the financial institutions (the Stock Exchanges specifically).

In the book, NNT defines the Black Swan events (BSe) as events that have three particular characteristics:

a. They are unpredictable;
b. They have huge (usually bad) consequences when they happen; and
c. Analyzing the BSe later (in hindsight), we realize that they were very much logical to happen.

NNT chose the name Black Swan, pointing to the surprise experienced by the Dutch sailors who, in 1697, saw the first black swan[2] near Perth, Western Australia. Before that, everybody knew that the swans are white.

NNT's book is mostly a philosophical book that deals with the reasons why we cannot predict the BSe. NNT is an "empirical sceptic" (as he likes to explain himself), ex-trader (New York, London, and Chicago) and ex-Dean's Professor in the Sciences of Uncertainty at the University of Massachusetts at Amherst. He is the founder of Empirica Capital LLC and a fellow at the Courant Institute of Mathematical Sciences of New York University. He mostly lives in New York (USA) and is, very much, a non-conventional person.

There are things on which I do agree with NNT and there are things with which I do disagree, mostly from the philosophical and practical point of view. As a simple example of my disagreements with NNT, I can present the "narrative fallacy". He strongly criticizes the "narrative fallacy" in the book, but I found it very useful in my teaching classes as one of the best ways to transfer the knowledge from the teacher to the students. This is something that is connected with "the context of the things": The "narrative fallacy" is (maybe) bad for science (it can "mask" the meaning of the facts or results), but I found it excellent for education (if you know what you are doing)!

What is a "narrative fallacy"?

NNT has provided a Glossary at the end of his book and there the definition of "narrative fallacy" is presented as a human's need to find and fit a similar story or similar pattern to a series of (connected or disconnected) facts that we encounter for the first time. Except in the education, the "narrative fallacy" is very much used in the statistical applications of data mining.

---

[1] In this book, I will use the acronym NNT as a reference to Mr. Nassim Nicholas Taleb.
[2] Later, the scientist gave it a scientific name: Cygnus Astratus.

DOI: 10.1201/9781003230298-2

If you go to the Internet, you will find other definitions, but most of them mean the same thing: We "innovate" or fit a well-known (logical, but subjective) story to the facts we have in front of us. We do that because it helps us to understand. Not always is the story good to fit the facts, but to us, it makes a sense.

In general, I appreciate NNT very much for his efforts against the (so-called) "orthodox scientists" who (from one side) "mask" the point of science and hide themselves behind elitism. These are PhDs with ordinary (and obsolete) ideas and methods of scientific research. Put another way, they cannot recognize good ideas, simply because these new ideas can "destroy" their understanding of the world. Anyway, I do not always appreciate the methods which he is using against the elitism in the science. I do not like to make the impression that I am against NNT's statements, but my purpose and my approach are different. Nevertheless, I strongly recommend all to read NNT's book!

There are plenty of statisticians and traders who agree and disagree with NNT's Black Swans and the explanations provided about them. Although NNT provides very strong arguments for the BSe in the Stock Exchange and in the life, most of the statisticians and the traders attack his book due to the lack of a more solid scientific basis to prove his hypothesis on the historical data samples.

The point of this book is not to debate regarding Black Swans. I will not strongly "stick" to the content of NNT's book, but I will "borrow" the three characteristics of the BSe and I will consider them from the point of view of safety in the Risky Industries.[3] Whenever I use something from the book, it is not with intention to analyze it, but only with intention to explain the impact of the BSe in the safety area in the companies from the Risky Industries!

2.2 CAN THE BSe Be CONSIDERED IN THE RISKY INDUSTRY?

The question in the title of this paragraph is a very basic question!

In my humble opinion, the answer is: It should be considered, but "the context of the things" cannot be same as it is in the Stock Exchange area!

NNT is speaking in his book mostly about BSe from the point of the financial and the Stock Exchange operations, and this is an area considerably different from the area of the Risky Industries. The Stock Exchange activities are based on undertaking the risk and, there, science is (sometimes wrongly) implemented as a step to justify the risks. But in the Risky Industry areas, the most important activities are to eliminate or mitigate the safety risks as much as possible. There, the science is implemented to recognize and calculate the risks and to eliminate and/or mitigate consequences, if the risks materialize.

So, there is a totally different "context of the things" between the Stock Exchange and Risky Industry operations, as presented in Figure 2.1.

Figure 2.1 presents all events (as probability distribution) that can happen in our lives (or in the industry). The distribution is Normal (Gaussian), but it can be any

[3] Under the name of Risky Industries, I mean the industries like nuclear, petroleum and oil, chemical, aviation, etc. These are industries where the faults or failures of operations may have huge bad consequences for the lives of humans and damage to the environment and assets.

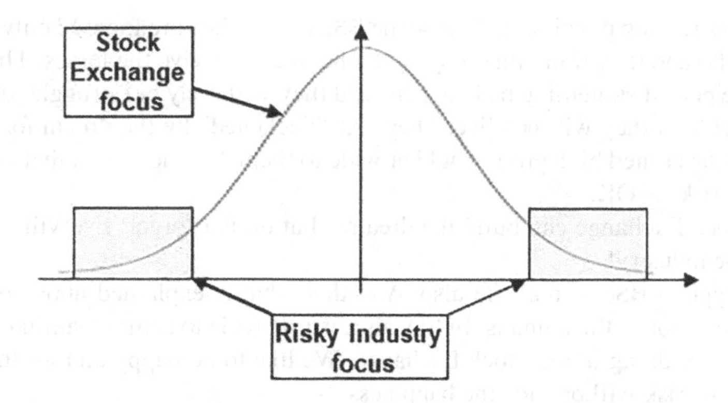

FIGURE 2.1 The different focuses of interest in Stock Exchange and in Risky Industries.

other curve (Cauchy, Student's-t, Weibull, etc.).[4] The point is that this curve explains that the good things (events) in our lives are present in the middle (so-called shoulders, close to *y*-axis[5]), which is quite normal. We strive to make our lives as much better as we can, so we try to increase the good events.

But, unfortunately, not always are the events in our lives good. There are also adverse events that happen to us, but they do not happen very often. It means that they are in the margins (tails) of the curve, left and right. These are areas marked with "Risk Industry focus" in Figure 2.1.

In the Stock Exchange area, the focus of operations is in the middle of the curve because everybody there looks for a profit (good things to happen). In this search for profit, people are selfish and they very often take risks to increase the profit (big risk – bigger profit). So, they assume that the events which will happen are in the area of the middle of the curve and they (intentionally or triggered by "narrative fallacy") neglect the areas on the margins of the curve (left and right tails) where the "adverse events" (BSe) are.

In the Stock Exchange, there are three types of employees:

a. The ones who prefer high risk, associated with high gain, are called Risk Loving (Risk Seeking) persons;
b. The ones who prefer low risk associated with low gain are called Risk Averse (Risk Avoiding); and
c. The third type are those who do not have rules how to behave and in some case they will avoid the risk and in, some case, will seek the risk. They are Risk Neutral.

---

[4] Having in mind that NNT is strongly against the use of normal (Gaussian) distribution, choosing it does not mean that I am provoking: The choice of the curve does not matter and there is no hidden agenda behind this choice! The Normal (Gaussian) distribution is chosen because it is available as a graph and only with intention to explain areas of interest of the Stock Exchange and the Risky Industry!

[5] On the y-axis are presented the probabilities of events and on the x-axis, the 0 is "good things happen". Going further left or right from 0 on x-axis the "goodness" of events is decreasing, so in the areas marked with "Risk Industry focus" are areas where "bad things" happen.

The Risk Loving people could cause the BSe there. They are focused only on their gains (profit) and they think that they (and others) can survive the losses. They (may be) will be poor if something bad happens and they will (may be) struggle for money later, but at least they will be alive. They are "blackened" by the dream for a better life which the earned high profit would provide to them. So, they think that accepting the higher risks is OK.

The Stock Exchange can build the dreams, but do not forget: Everything else is built by the industry!

NNT applied BSe to real life also. And that what is explained above is in line with the behavior of the humans: In life, also, we struggle to achieve similar gains as the traders are doing in the Stock Exchange. We like to be happy and we think that accepting the risk will provide the happiness.

Although the Risky Industry is taking care of the profit, they pay considerably more attention to the risks (adverse events) of their operations (marked areas with "Risk Industry focus" on the curve in Figure 2.1). If bad things happen there, the people will not lose just the money, but they and their families could lose (or ruin) their lives from the consequences of these wrongly accepted risks.

2.3 CONSIDERATION OF THE BSe IN THE RISKY INDUSTRIES

To explain the BSe in line with their presence in the Risky industries, I will stick to the three items of the definition of the BSe stated by NNT in his book and mentioned by me at the beginning of Section 2.1 of this book. The main question here is that, if we look at the three characteristics of the BSe, can we say that every accident in Risky Industry is actually BSe?

NNT's book has a significant philosophical aspect, but I will not go there. There are a lot of critics regarding the philosophical and the logical inconsistencies in NNT's book, and I will mention just few of them:

a. The subtitle of the book is "The impact of highly improbable". Using word "improbable" is not the same as "never happen before" or "unimaginable";
b. In few examples of the BSe, especially in the story of Lebanon's civil war, there is the impression that categorization of the BSe highly depends on the attitude of the observer;
c. All three parts of the definition seem not to be "scientifically" appropriate or with other words: Do not provide arguments for a clear definition of what is a BSe;
d. He uses the wording: Black Swan "lies outside of the realm of regular expectations.", but what is "regular expectation"? Can it be more precisely quantified?
e. Etc.

All those things will not be considered in this book. I will try to be pragmatic and apply "safety understanding" of the BSe (whatever that means…).

There is another thing about the book that needs to be mentioned here: NNT is using also the term Gray Swan events (GSe) and he states the difference between the BSe and GSe in the Table 1 in the paragraph The Tyranny of the Accident in NNTs book. The GSe, by NNT, are events that are "tractable scientifically" and BSe are those that are "totally intractable". In my humble understanding, from a scientific point of view, the GSe are those whose existence can be predicted by science, but has never been discovered before. As an example, I can mention the Higgs boson[6] is GSe: Its existence was theoretically predicted by Peter Higgs (and five other scientists) in 1964, and it was discovered in CERN in 2012.

I will focus on the GSe in this book later, so let's continue providing more details through all three parts of the definition of BSe…

2.3.1 It Is Unpredictable…

Every incident or accident in the operations of Risky Industries is unpredictable!

Someone will say: But we produce the List of Hazards and we calculate the risks where the probability for each hazard is assigned, so how it can be unpredictable?

I can answer this question with another question: If it is predictable, why can we not stop it? Maybe sometimes we can predict what will happen, but usually it is too late: There is no time to stop it…

But the things are not so simple… The point with the question from above (regarding the List of Hazards) is hidden in the word "probability".

Having probability that something would happen does not mean that it is predictable. Predictability, as a word, is strongly connected by Determinism, while probability is a word connected with Randomness and uncertainty. In Determinism, it is clear: Following the equations and laws, we can predict with 100% that something will happen. Theoretically and scientifically, having 100% probability is actually not having Randomness and uncertainty: It is Determinism.

But (again), the things are not so simple with probability… There is always more…

As we know, probability can be expressed by numbers between 0 and 1 (or by percentage from 0% to 100%). Having 0 (0%) probability, scientifically (deterministically), means that this event will never happen. And also, having 1 (100%) probability, scientifically (deterministically), means that this event will always happen.

Either way, in the real life, even if we have 0 (0%) probability, the event may happen. and even if we have 1 (100%) probability, the event may not happen. These situations can be explained as extremely rare events that do not fit the theory, but they are allowed in practice. In addition, if there is probability of 30% of some outcome of experiment to occur, it does not mean that if you try the event 10 times, this outcome will always happen 3 times. It actually means that if you try 1,000,000 times, the outcome will happen **approximately** 300,000 times! As you can see, uncertainty is always present with the events and outcomes based on probability.

The statements above are the core of understanding the probability: These are just mathematical calculations that do not mean that they will be fulfilled always in real life! In the case of tossing a coin, we assume (after many experiments) that the probability will be ½ for "head" and ½ for "tail". But these numbers are just hypothetical

[6] Higgs boson is elementary particle which, used by Higgs mechanism, explains why particles have a mass.

numbers that (theoretically) will show up when we toss the coin an infinite number of times. Mathematically, it can be expressed as:

$$P_t = P_h = \lim_{n\to\infty} \frac{k_t}{n} = \lim_{n\to\infty} \frac{k_h}{n} = \frac{1}{2}$$

where P_t (P_h) is the probability of "tails" ("heads"), k_t (k_h) is the number of "tails" ("heads") in a tossing experiment, and n is the number of times the experiment is repeated (which theoretically should be close to infinite).

In general, although the probability is strongly connected to the Randomness and it is full of uncertainty, we should be careful. The real-life definition of the Randomness is not easy because the randomly determined events can give perception of outcomes that do not look random.

I can give the next example: Random outputs also mean that, in the case of many variables, they can adjust (randomly) themselves as deterministically. Having a series of outcomes as 1, 2, 3, 4, 5, and 6 while consecutively tossing the dice 6 times , the outcomes look extremely deterministic. If after the fifth toss, someone asks what will be the next number on the sixth tossing of the dice, having in mind the first 5 tosses, everyone will answer: 6. But in reality, tossing a dice is a random operation, and each outcome is independent (not connected with previous ones), so the probability of the sixth toss of the dice to be 6 is only 1/6, in any case. It does not mean that sixth toss to be 6 will not happen, but the probability that any other number will be the outcome is the same as the probability of 6 to be the outcome!

The reverse also applies: The deterministic events can generate outcomes that look totally random. So, there is a need to be cautious when you try to determine the Randomness or Determinism of the operation or the process.

2.3.1.1 Extension of the Definition of Probability

To clarify the "extension of definition of probability", I would use an example with three types of the hazards that can show up in the Risky Industries.

The first ones are known hazards. These are hazards that are experienced in the past and will be part of our List of Hazards. These are hazards for which the data are available and we can transform them into risk by calculating their frequency[7] (how often they can happen) and their severity (how strong will be the consequences). The overall situation with these hazards is well known, and there are already methods and activities on how to eliminate or mitigate the risks of these hazards.

The second type are hazards that can emerge (just pop up) during our daily operations (processes, activities, etc.). These could be hazards that are on our List of Hazards or they may not be on the list, simply because they do not happen very often, so they have been forgotten.

Having in mind the dynamical environment that is present in the Risky Industries, the need for monitoring our operations is a very important regulatory and operational requirement. The monitoring will help to notice and recognize the hazards and risks that are known and unknown, so appropriate action will be triggered. There

[7] Or likelihood or probability...

is another aspect of these emerging hazards: They can even happen when we intentionally change our process (operation, equipment, employee, etc.). Each of these changes of the process could eliminate or mitigate some of the risks, but they can also create the new ones which (maybe) are not in our List of Hazards. That is the reason to reinvestigate all hazards and recalculate the risks, before and after the change take place.

The third type of hazards are those which we do not know or do not assume that they exist. This would be the type of hazards known as BSe. Not knowing that these are possible make us unprepared to fight them. Actually, we shall be prepared even for such a situation, but let's assume that they will be pure BSe.

To continue with clarification, I will go in the areas of probabilities. Speaking about connection between the probability and the risk, the probability deals only with the frequency (likelihood) of the risk. For the sake of truth, one hazard may result in many different consequences, but how often any of these consequences will show up is subject to probability.

Let's say that we have an operation where many adverse events can happen and the possible number of all outcomes (adverse events) is k. It means that the all outcomes of the adverse events are set with k different elements, or, in other words: k is number of different types of incidents and accidents. The probability of the occurrence of all the outcomes of the adverse events (incidents and accidents) can be stated by the equation:

$$P(\text{adverse event}) = \sum_{n=1}^{k} p_k = 1$$

where $P_{(\text{adverse event})}$[8] is the probability of any incident or accident to happen and p_k denotes the probabilities of different types of incidents or accidents which may happen.

As it can be noticed, there are different types of adverse events that may happen, and all of them (together) are producing probability of 1 (100%). Be careful: This is not a probability that any of these adverse events will happen in the scope of our operation (process, system, etc.). This is just a total probability of all our known (registered) adverse events. In other words: If any of the adverse events would happen, it must belong to the $P_{(\text{adverse event})}$ set. Inside the $P_{(\text{adverse event})}$, each of the adverse events could have different probabilities of occurring (some will happen rarely and some more often), but the total probability that any of them could happen (as adverse event) will be equal to 1 (100%). In the equation above, the risks for hazards of the first type and some of the hazards from the second type will be included.

Bearing in mind that I have been in the safety field since 2005, once I heard that "...the probability something bad to happen is the highest for the adverse events which we never assumed that it may happen...". These would be, actually, a risk for hazards from third type and some of the hazards of second type.

[8] In probabilistic theory, the $P_{(\text{adverse event})}$ is known as the "sample space" This is actually a set of all possible adverse events which are known and they could happen in a particular experiment (in our case: In particular company).

Considering that, the abovementioned probability equation could be changed into:

$$P(\text{adverse event}) = \left(\sum_{n=1}^{k} p_k \right) + P_{\text{NA}} = 1$$

where the P_{NA} is the probability of the outcomes of the events that were "Not Known or "Not Assumed" that could happen.

Now, going back to Section 2.3.1 (It Is Unpredictable…) I can say that actually, P_{NA} is reason why I have written there the following sentence:

Anyway, in the real life even if we have 0 (0%) probability, the event may happen and even if we have 1 (100%) probability, the event may not happen.

In other words, the P_{NA} is the probability (in "the context of the things" with present understanding of safety probability) that the hazards which are not in our List of Hazards will materialize as a risk and later as an incident/accident (BSe). If, during making the List of Hazards, some of the hazards were not assumed that can happen, there is no chance to calculate the probability or severity of consequences for them, or in simple words: No hazard registered – no risk calculated!

In our case (the "revised" equation above), the P_{NA} is actually a contribution of the BSe (P_{BS}) to the overall probability $P_{(\text{adverse event})}$. That is,

$$P_{\text{BS}} = P_{\text{NA}}$$

There is one of the famous Murphy's Laws that can be associated with the BSe:

If you have predicted that there are four possible ways in which an operation can fail, and you have provided prevention to overcome each of them, then a fifth way, unprepared for, will develop, for sure!

Simply, the BSe can happen in reality.

As a simple example, such an outcome (element in the set of all terroristic attacks in the history of humanity) is the 9/11 attack that occurred in the USA. Before that, there was no such outcome, so nobody assumed it would happen. It is clear that the 9/11 attacks belong to the P_{NA}.

As an additional (real-life) example, please note that, in the List of Hazards of the airlines, flying over East Asia must be aware of the hazard of volcano eruption. Volcano eruptions in these areas are common events that release huge clouds of ash, which can cause the flying aircraft to crash. But having this hazard on the List of Hazards in the airlines which are flying only over Africa[9] will bring so small a risk in regard to probability that we can assume it is zero probability.

In general, the P_{NA} means that there are things that we do not know. We do not know that they are possible in our environment (our "context of the things") and/or we do not know that they exist.[10]

[9] Maybe also for Europe, except the Iceland and maybe Italy…

[10] In this sentence you can find the core of the Black Swans: As it has been said before, until year 1697 no one knows that black swans exist, simply because no one has seen it…

So, the second (revised) equation in this paragraph is more mathematically correct than the first one for the area of Safety. Anyway, having in mind "the context of the things",[11] it is not applicable in real life. The problem with this equation is that P_{NA} will also include the adverse events that do happen due to the wrong assumptions established in the cases with high uncertainty (expressed by σ)[12] regarding the bad outcomes of particular good event. This is something which, in the industry, is known as choosing "too wide tolerances", where the product (service) cannot any more provide a safe use.

2.3.1.2 Statistics and Probability[13]

To calculate the probability, we need statistics, and to implement statistics we need data. But appropriate data is not always available, so depending on the data available, there are two types of statistics. The first one is exclusively based on all available data and it is called **Descriptive statistics**. This is a statistic which describes only the set of data through the calculations of average, standard deviation, skewness, kurtosis, etc.

The second type of statistics is **Inferential statistics.** This statistic takes samples of data from the set (population) and the set is usually a huge amount of data. By processing these samples, it can infer about the characteristics of all data in the set (population). The Inferential statistics can be divided into two catcgories: **Frequentist**[14] and **Bayesian.**

The Frequentist statistics uses experiments and data gathered from the experiments. It is mostly used when there is random process. Simple examples are tossing a coin or dice: You toss a coin or a dice 100,000 times and the frequency of "head/tail" or numbers on dicc will be presented through following equation:

$$P(k) = \frac{n_k}{N}$$

where the $P(k)$ is probability event k ("head" / "tail" or any number on dice) to happen during normal operations; n_k is number of event k happening; and N is total number of normal tosses (in our case 100,000).[15]

Very often, we are not in a position to do experiments (or maybe doing experiments is too expensive and time-consuming process), so we may use the Bayesian statistics to provide the probability. The Bayesian statistics uses the association between calculated data and previously gathered knowledge regarding some events.

A simple example for the Bayesian statistics is the next-week football match between two teams during national league matches. To provide the probability of outcome of the match (win/draw/lose) for each of the teams, wc can use the statistics of the matches from the previous years, which is well documented. But if one of the

---

[11] Explained in Section 1.4 of the Introduction!

[12] Will be explained later in the book!

[13] Please note that I have not found the introductory book in statistics which is less than 250 pages and the introductory book in probability which is less than 400 pages. So, whatever is written in this paragraph and in this book is just small part of these two mathematical disciplines.

[14] Maybe, not everyone will be happy with this name, but I found it somewhere and I like it!

[15] In the theory of probability, this way of calculating probability is known as Basic Counting Principle.

teams has changed the manager, it is clear that they will probably change also the style of playing. If there are also changes in the players, we can agree that previous years' matches are not so much of use. This is the simple case when the previous data has nothing to do with the present situation. In such cases, we need to look at the present situation on the table and at the new scores with the already played matches with the new manager and the new players. Any other sources of information could also provide data for the Bayesian statistics.

Bayesian statistics is a very powerful tool because it can take care of the changes in the system (new manager or new equipment). The problem is that we need to be cautious with that because the Bayesian statistics can be subjective. As already mentioned, it is based on previously gathered knowledge regarding the event of interest. The point is that the understanding of this event of interest (knowledge) can vary between the analysts. Simply, different analysts will have different conclusions on how to use the same knowledge. The issue with Bayesian statistics is that it can vary in time also: Simply, by investigating the previous history and the present state of the situation (as the investigation proceeds), we gather different knowledge. Using this different knowledge, it could provide different results.

In the Risky Industries, to calculate probability, both of the statistical methods are used, but which one will be used depends on the data availability. Mostly, due to general lack of data, instead of the word "probability", the word "likelihood" is used. Scientifically, I think that more convenient word to be used is "frequency" instead the "likelihood", but you may decide yourself what to use.

The difference between "probability" and "likelihood" is in the value of N in the equation above. If the N is extremely big (let's say few millions, ideally infinite), then the likelihood becomes the probability. In the Risky Industries, the adverse events do not happen very often (n_k is small!) and the operations are not infinite in number (although N is huge), so the data for calculating probabilities for adverse events are missing. In the area of the Stock Exchange, the data are abundant, so the experts there need to use probability distributions. I do not say that this is good or bad for the Risky Industry, but it is the pure reality (unfortunately).

And now, at the end of this paragraph, let's clarify something regarding probability. For the sake of true, probability cannot be connected only with the experiments or events. In addition, we use probability also to describe our beliefs regarding some outcome of some event (activity, operation, process, etc.) and also to express the chances about some outcome during betting. So, as you can understand, "the context of the things" regarding the probability cannot be expressed only by experiments used in science, which means that you should extend the definition to other types of analysis.

2.3.1.3 Probability Distributions

Speaking about probability, we cannot neglect the probability distributions...

The probability distributions are mathematical functions that gives a rule on how the outcomes of set of random variables can show up when the particular experiment (operation, process, toss a coin, etc.) is executed. They are very important because they are used to, theoretically, calculate the probability of a particular outcome.

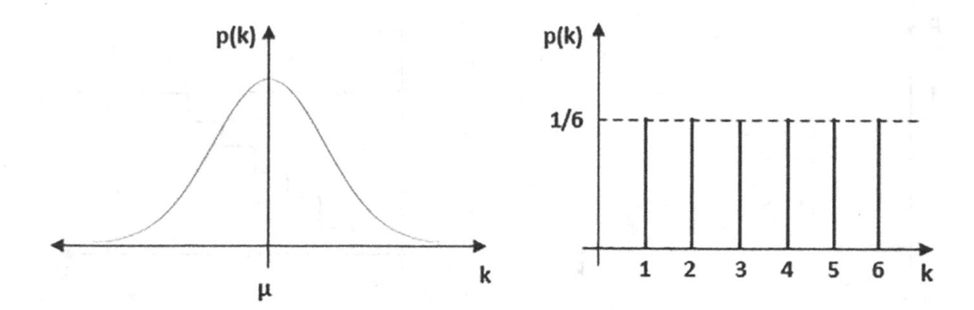

FIGURE 2.2 Continuous probability distribution (Gaussian, left) and discrete probability distribution (tossing a dice, right).

The probability distributions (known also as probability functions[16]) can be continuous or discrete (Figure 2.2).

Calculating probability by the use of probability distributions for particular events which are discrete is easy: You just look at the event on the x-axis and you read probability on the y-axis. With continuous probability distributions, the things are a little bit complicated. With continuous probability distributions, each dot from the curve does not represent an event. If it does not represent a single event, then the probability that any dot "will happen" is equal to 0. It means that value of each dot on the y-axis is not a measure of probability.

The probabilities (as numbers) for continuous probability distributions are calculated as interval in the limits of integral. So, if we would like to calculate the probability (as number) that the experiment will results with an outcome equal to and smaller than 4, then the equation for calculating will be given by following two equations:

$$P(4) = \int_{-\infty}^{4} p(i)\, di \quad P(4) = \sum_{i=1}^{4} p(i)$$

where $P(4)$ is probability outcome $k=4$ and less than 4 to happen by an experiment. The left-hand side equation is for continuous situations (outcomes), while the right is for discrete situations (outcomes). Actually, this is a definition of cumulative probability distribution.

Graphically, the cumulative probability distributions (functions) for the probability distributions from Figure 2.2 are presented in Figure 2.3.

The important characteristics to the cumulative probability functions are given by these equations:

$$P = \int_{-\infty}^{\infty} p(k)\, dk = 1 \quad P = \sum_{k=1}^{n} p(k) = 1$$

[16] In the literature you can find also the name "probability mass function" for discrete values and the name "probability density function" for continuous values.

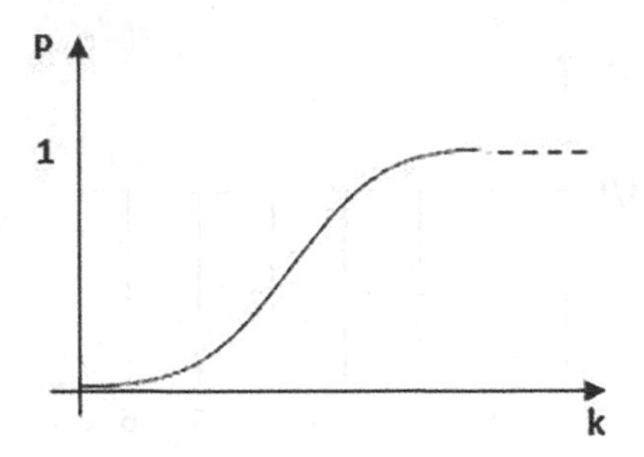

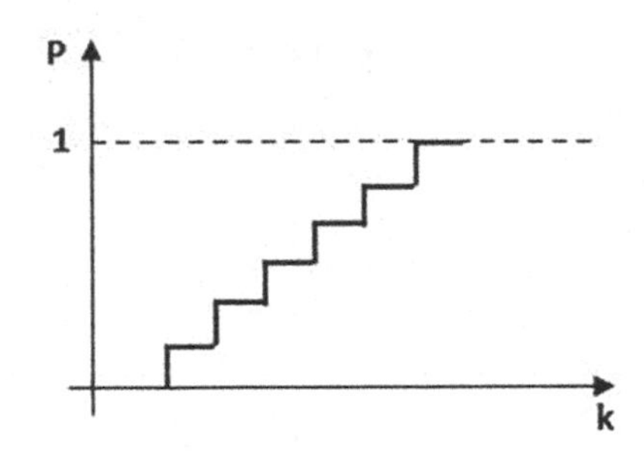

FIGURE 2.3 Cumulative probability functions for continuous probability (Gaussian, left) and for discrete probability (tossing a dice, right).

If we would like to "insert" P_{NA} in above equations, the equations will transform into:

$$P_{tot} = \int_{-\infty}^{\infty} p(k)\,dk = 1 - P_{NA} \quad P_{tot} = \sum_{k=1}^{n} p(k) = 1 - P_{NA}$$

As you can notice, changing the equations for the cumulative probability distributions by introducing P_{NA}, would change the basic rule in mathematics that the sum (integral) of cumulative probability distribution functions will be 1. Of course, it will be valid only for the applications in the safety areas!

So, it is important to understand that the probabilities that any event to happen are numbers and the probability distributions have all possible outcomes from the experiment or set of events which could happen, if some operation is executed. The probability distributions could be used as a tool to calculate any probability that some or few events will happen from the set of events that could happen.

That is the reason that NNT considers the probability distributions: Assuming that particular set of data is belonging to wrong probability distributions means that calculation of probability for the adverse event will be wrong. It is not an issue if the calculated probability is bigger in the wrong calculation. For adverse events, the higher probability means that we will reconsider our decision to ignore this event. But if the calculated probability is lower in the wrong calculation, it will give us an additional "boost" to make a wrong decision.

Although the probability distributions and the cumulative probability functions are essentially connected to the probabilities, they are not used very often in the safety area. The reason is lack of data and the discrete nature of the adverse events in the safety areas. The probability distributions have their purpose mostly in continuous systems of data.

Lack of data is lack of knowledge regarding the BSe, and this would affect predictability of the BSe in the Stock Exchange and in the Risky Industry. But regarding the unpredictability, the situation with the BSe in the Risky Industries is different from that in the Stock Exchange. I can state that the unpredictability in the Risky Industries is caused by lack of data, but in the Stock Exchange they are caused by the wrong assumptions regarding the probability distributions connected with the data.

There, and in our normal life (as NNT explains in his book), the people mostly rely on the Normal (Gaussian) distribution instead to others (so called "Fat-tail" distributions). The calculations regarding probabilities of the BSe based on the Normal (Gaussian) distribution produce significantly different results. Putting into consideration the assumption that your data abide to the Normal (Gaussian) distribution instead to the "Fat-tail" distribution means that you are underestimating risks for the adverse events. It happens because the tails of the "Fat-tail" distributions provide bigger risk than the tails of the Normal (Gaussian) distribution.

In general, the use of improper probability distribution will produce fundamentally wrong results.

This is something that can also affect the uncertainty of measurements (as it has been explained in Section 1.5.2). Assuming Normal (Gaussian) probability distribution will also change the value for standard deviation, so the overall probability where the true value is will be changed for any $\pm\sigma$ (or $\pm 2\sigma$ or $\pm 3\sigma$) areas.

This is another reason why the BSe cannot be analyzed in the Risky Industry in the same way as they can be analyzed in the Stock Exchange area. Anyway, the considerable level of unpredictability is present even in the Risky Industries...

In general, whenever we deal with data, we use statistics. But to produce some plausible results regarding probability for the future events (based on present data by using statistics), we must find the proper probability distribution for the used data. Without it, whatever the result of the statistical analysis will be, the certainty of prediction will be very much questionable.

2.3.2 It Has Huge (Usually Bad!) Consequences When It Happens...

The BSe does not necessarily have only bad consequences!

They can happen also on the "other side of the road": Winning a lottery is a BSe, but it is a good BSe! Internet was not predicted, but it did occupy our attention and our lives; without it, our lives would be terrible! Do not think that the BSe are only adverse events!

I always thought that parents need to teach the children how to deal with the bad things in the life. For the good things, children will not need education. So, having in mind that this is a book for safety, I will deal here only with the bad BSe.

Any adverse event which happens produces bad consequences! That is the reason that we call it an "adverse event". And these bad consequences are the reason we do not like the adverse events. Actually, this is the definition of adverse events: Events, which produce unpleasant (bad) consequences when they happen to us.

The adverse events in the financial world are mostly market (Stock Exchange) crash or bankruptcy of the company. If such a thing happens, there are consequences: Many people are fired, they lose their jobs and their incomes and savings, and prices go down, but the big wisdom in economy is: Sell when the prices are high and buy when the prices are low! I have a friend who was lucky to have money in a safe place where the house-triggered crisis in USA (defined as BSe from NNT) started in 2008. He used the significant decrease of the prices of the houses in USA and bought a very cheap house for his family. Under normal conditions, he could only dream about such a house. So, however the BSe is bad for the community, somebody will be lucky...

In the Risky Industries (as mentioned before), there is categorization of the adverse events into incidents and accidents. In simple words, the incident is an event where someone has died or is hurt and/or there are some damages of the assets, but these damages can be fixed. The accident is an event when many have died and/or the damage of the assets is catastrophic (they cannot be repaired and/or used anymore).

As we can notice, the incidents and accidents differ by the level of consequences. The incidents are with consequences which can be healed (repaired) and they will not produce catastrophic change of our lives, while the accidents are with consequences that can totally destroy our lives (assets) or, at least, the recovery is not certain and not easy at all.

Having in mind NNT's definition of the BSe, we can say that, in the Risky Industries the incidents are not always BSe, but also not all accidents are BSe. By definition, only the accidents that are not in our List of Hazards can be defined as BSe. The catastrophic consequences would be death of many people and damage of the equipment beyond repair, and this is appropriate with BSe.

Having this in mind, there is a very logical question about the incidents: Should we dismiss the incidents in the Risky Industries?

The answer is clear: Not at all!

The incidents have an important role in the safety management, and their role is that they are powerful indicators that something is wrong in the company. If the preventive or corrective measures are missing after any incident, then the company is "ready" for accidents. There is another thing regarding the incidents: If the incident is not stopped on time, it could transform itself into accident. That is the reason that they should not be neglected.

Speaking about that, there is a question which will pop up: Is the company with many incidents and no accident a safe company? In such a company, the accident is hidden and could happen any time. This is the reason that, registering many incidents in the company in the Risky Industry, the State Regulator must react. The employees there must find the reasons why this happens but, in general, it could be, very much, a systematic error. If there is systematic error in the company, then it cannot be found through Internal Audit. That is the reason for periodical Regulatory Audit: It will provide a good chance to register it.

In regards to the incidents and the accidents, I can mention another difference between "the context of the things" in the Stock Exchange and in the Risky Industry. In the Stock Exchange, the "incidents" (small loses) are part of everyday life, but in the Risky Industries they are very powerful warning signals that something must be done.

Determining the consequences of the BSe (which are disastrous), especially in the area of the Stock Exchange, can be subjective and, as such, is questionable. Maybe the real question there is: Is there any real and clear criteria about the BSe in the Stock Exchange?

Having in mind that NNT is mostly dealing with global BSe, the question is: What happens with all these individual adverse events that happen to the individuals when they try to earn some money on the Stock Exchange? There are plenty of such individuals who are investing their or other's money; every day, plenty of them lose the money. Is that a BSe for their lives and families? They are simply neglected as BSe by NNT, regardless of the impact of the lives of these humans being catastrophic.

From another perspective, in the Risky Industry, every loss of life is BSe (collectively and individually).

This is one of the big differences between the BSe in the Risky Industries and the BSe in the Stock Exchange: In the Stock Exchange, the individual losses (individual catastrophes) are neglected, only the country or global (big) losses are counted. There, the individual losses are not signatures for possible general warnings as it is with the incidents and accidents in the Risky Industries.

In general, if we need to investigate the adverse events in the Risky Industries, as they have been defined in the Stock Exchange (or in our ordinary lives as it is explained by NNT), it will not work. We need to take different approach for that...

As a final statement I can say: In the Risky Industries, there are adverse events which are BSe and there are adverse events which are not BSe. Those accidents which are not BSe, we know that they can happen, we already have them in our List of Hazards. So, we try to implement corrective and/or preventive actions to eliminate or mitigate them.

But sometimes, we simply fail...

2.3.3 Analyzing the Event Later, We Realize that It Was Logical to Happen...

After every incident or accident (BSe), there is Regulatory requirement for investigation in the Risky Industries. There is also an obligation of reporting any incident/accident (also required by the regulation) and any case of neglecting this obligation is a subject of legal prosecution. In the Stock Exchange, there is no need to report anything: Everybody can see it, all around the country or all around the world. And there is no country investigation or global investigation for the BSe in the Stock Exchange required by the Regulation.

The investigation in the Risky Industries should provide data that can be used to stop the accident from happening again (Lesson Learnt). In the Risky Industries, the investigation goes as far away until the root-cause is not determined. There, the regulation for implementing Safety Management System with all its characteristics and requirements is multi-faceted. The most important outcome from the investigation must be: Is the root cause for the adverse event acute or systematic error?

The systematic error is more critical, and it can provide more measures that could affect not only the company but also the industry. From another point, a systematic error, once registered, can be easily eliminated in the entire industry by the change of the regulation.

The investigation must provide data based on integrity and objective evidence, which can be used to establish corrective and/or protective measures to eliminate the root cause and the consequences, for possible future cases of such event. If it is not possible to eliminate the future happenings, then efforts must be applied to try to mitigate the root cause and/or consequences. If even that will not work, emergency procedures, back-up plans, and contingency plans must be provided.

In addition, the investigation in the Risky Industries is needed to check if there were some unlawful activities of the subjects involved in the accident. The possibility of unlawful activities will be investigated also in the Stock Exchange area, but there the situation is quite different.

In general, all these requirements for the SMS and for investigations should provide explanations that are extremely logical and, normally, we should realize how the incident/accident had happened. But, as I said before, the situations with the BSe in the Risky Industries are quite different than situations in the Stock Exchange: In the Stock Exchange, there is no requirements for SMS! There is something else, but it is quite different then the SMS in the Risky Industries.

An important requirement that must be included in the SMS is: It must be proactive! It means that each subject must produce a List of Hazards and each hazard from the list must be determined by likelihood of happening and severity of consequences. If the Fault Tree Analysis (FTA) or Failure Mode and Effect Analysis (FMEA) is used in calculating the likelihood of the accident, then it shall check the results of investigations in accordance with the previous assumptions for the criteria.

In this area, the 9/11 event (used also by NNT in his book as example of BSe) is very important. Regarding this event, the aviation safety authorities (who were calculating probabilities for different terroristic attacks) did not have such an event in their data, simply because it never happened before and no one assumed that it could happen.[17]

So, recalling the second equation from Section 2.3.1.1 (Extension of the Definition of Probability):

$$P(\text{adverse event}) = \left(\sum_{n=1}^{k} p_k \right) + P_{\text{NA}} = 1$$

we can conclude that, by NNT and his book, the 9/11 attack (BSe!) belongs to the variable P_{NA} in the equation.

This situation (maybe) should be considered as existence of the BSe in the Risky Industries, but under this name should come the events (incidents and accidents) that were never assumed as ever happening, not because the likelihood for them was too low but due to the fact that no one assumed that such an event could exist.

In general, in each company in the Risky Industries, in their SMS Manual, there should be a procedure for Hazard Identification (to produce a List of Hazards!). This is mostly done by a brainstorming session of different profiles of employees in the company. During these sessions, these employees list different hazards that are specific for their area of expertise and their working positions in the company. Registering the event that can be attributed as a BSe later means that it will belong to P_{NA}. In other words: During these brainstorming sessions, no one listed such an event as hazard. It may look strange, but it is extremely normal. "Humans are strange animals" is an expression which I am using often when I would like to explain the

[17] Although, this event never happened before, the FAA reacted very fast and very wise: They closed the airspace over USA. The aircraft which stayed on the primary radar was actually the terroristic aircraft. It helped the fourth hijacked aircraft (United Airlines flight 93) to crash in Pennsylvania.

situations that were beyond every expectation of the behavior of some person. This is something that could be connected by Human Factors (HF), and I will consider them later in the book.

The main common point regarding "the afterwards realization that the BSE could be predicted" in the Stock Exchange and in the Risky Industry areas is that (I agree with NNT) we use "narrative fallacy" to assure ourselves that we should have predicted this event.

And this is something which is worth paying attention to from the point of view of psychology of humans and HFs also. We (as humans) possess the DNA where our expectations (based on our past life and professional experience) are embedded. This is something which is hard to change, and this is maybe one of the reasons that our life or professional experience has importance on the same level as our gathered formal educational knowledge. The synergy between the educational knowledge, the skills gathered during our lives, and the professional experience is one of the most important tools for building an attitude for fighting the BSe in real life.

2.4 THE CHARACTERISTICS OF THE BSe IN STOCK EXCHANGE AREA

To fight the BSe, we need to get familiar with them. We need knowledge about BSe, and this can be obtained from the data. We need to gather particular amount of data, to analyze the data, and to get conclusions about the BSe.

The conclusions would trigger particular decisions. In the Stock Exchange, the statistical models are based on possible outcomes that almost always limit the events assumed to be outliers in the gathered data. In the Risky Industries, this simply does not apply.

The explanation here (in this paragraph) is a simplification and, as such, it should be taken with understanding that, in general, the overall nature and flow of data and activities in the Stock Exchange areas are, pretty much, complex.

In the Stock Exchange, data is gathered about movement of the indices every moment, and all these data are analyzed as whole. "Analyzed as whole" means there is no difference between "ups" and "downs" of the indices. "Ups" (good things) are part of the data and "downs" (bad things) are also part of the data. Having in mind that activities on the Stock Exchange are based on human assumptions, it is understandable that the humans are taking care to gain the profit from these "ups" and "downs". For some companies, "ups" are good (they can earn money by selling) and for some companies, "downs" are good (they can earn money by cheaper buying). It is very much a complex and unpredictable movement, and I would not like to go into details here. But roughly and simply speaking, let's say that number of "ups" indices will be good things and, as such, their number will be considerably bigger than number of "down" indices (bad things).[18] In such a case, the humans are striving to produce more "ups," and it is understandable that the numbers will have such a ratio.

[18] These "ups" (good things!) can be data from days when the markets are going up and "downs" can be data from days when markets are going down. If the market is going abruptly or better to say catastrophic down, it is BSe!

The point, in regards the BSe in the Stock Exchange, is that there will be approximately few times more "ups" than "downs". The amount by which "ups" will go up will be small. With "downs," this is different: The number of the "downs" will be smaller, but when they happen, they will go down for bigger amount than "ups" go up. The panic when this will happen will increase the speed and amount of going down of indices like an avalanche effect, and thus BSe happens!

Why it could happen?

Because in the Stock Exchange are many traders in the same room and they share the same information. If some of the more experienced traders start to sell commodities when they noticed that the indices are going down, there is a very high probability that the not-so-experienced traders will follow suit. The huge selling will produce decreasing of the price, and this is a BSe for that company (or even for the industry). This is something which is known as "social phenomenon" and, as such, it may not be presented by Normal (Gaussian) distribution.

In the Risky Industries, such a thing cannot happen. There, every activity of one company is analyzed from the "the context of the things" of the other company. Whatever happen to one nuclear plant does not necessarily mean that it will immediately happen to another nuclear plant.

In the Stock Exchange area, there are financial models based on data, and they are used to predict the "next move" on the markets. In most of these models, the pillar is Normal (Gaussian) distribution (Modern Portfolio Theory, Black-Scholes models, Capital Asset Pricing model, etc.). There, it is very often when the outliers are simply neglected (discarded from other data). The theory there is mostly based on the Normal (Gaussian) distribution for practical reasons: This distribution can be easily described using only two variables: The mean (μ) and standard deviation (σ). It is common in the Normal (Gaussian) distribution to neglect the outliers: They simply do not fit into the rule. The "Fat-tails" distributions are more sensitive to the frequent small changes and to the infrequent big changes and, as such, these distributions fit better in the Stock Exchange data.

The point is that all data regarding "ups" and "downs" will produce a model with some probabilistic distribution, and the "ups" will be close to the optimum (to the mean, μ) and "downs" away from the optimum. Having in mind this, the BSe in the Stock Exchange are defined as events that happen not very often but have catastrophic consequences when they do, we can assume (very accurately) that the BSe will be more far away than ordinary "downs". As such, they will be "outliers,"[19] and they will belong to the far ends of the tails of the distributions. In Figure 2.4 are presented places in probability distribution where good data is (shoulders!) and where the bad data is (tails).

If you read this book with due attention, you will notice that Figure 2.4 is the same as Figure 2.1. But please bear in mind that "the context of the things" of these figures is different. Both figures are used to explain different things. Figure 2.1 presents areas of interest of the Stock Exchange and the Risky Industries and Figure 2.4 explains "shoulders" and "tails" of the probability distributions, which are very much important for the area of Stock Exchange. It means the data from the "good" days on the Stock Exchange are usually presented together with the data from the days when the BSe happens.

[19] Outliers in statistics are defined as points which significantly differ from other data and as such these are extremes which can be found on the tails of the distributions.

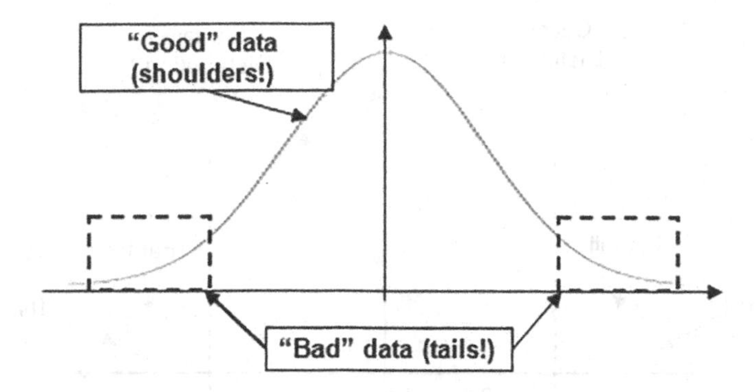

FIGURE 2.4 The places in probability distributions of good (shoulders) and bad data (tails) in the Stock Exchange.

The problem with the BSe in the Stock Exchange is that most of the guys there are using the Normal (Gaussian) distribution to calculate probabilities of the BSe. And this is wrong…

Why is it wrong?

Because, as mentioned few times: "The context of the things" in the Stock Exchange.

There, people try to earn money and each rise of the market will bring them benefit. But the point is that they need to predict the rise of the market and, at the same time, calculate the losses, if the rise does not happen. In this calculation, the use of Normal (Gaussian) distribution will calculate lower risks than if "Fat-tail" distribution is used, simply because the gain is in the "shoulders" and the loses are in the "tails" (Figure 2.4).

It is wrong to use Normal (Gaussian) distribution because there are other distributions that better fits the data from the Stock Exchange. These distributions (see Figure 2.5) have "fat tails" (bigger than Gaussian) and as such they produce bigger probabilities for the BSe to happen.

In general, using data as Normal (Gaussian) distribution instead as "Fat-tails" distribution, we underestimate the probabilities about the BSe. It means that we assume that risk (frequency of the BSe) is smaller and it will not materialize itself so often. Starting from this position, it is clear that, when it happens (and it will happen more often that we had assumed), we will be very disappointed. This is the case when the BSe in the Stock Exchange will happen.

The problem with the investigation of the "Fat-tails" distributions[20] is that there is no one unique definition. On the contrary, depending on "the context of the things", there are plenty of different definitions for these. The additional problem with all these different definitions is that they provide different points where the "tails" starts to be a "tails". One if the most used definitions is a definition that the "tails" start on a distance far away than $\pm 2\sigma$ from the mean (μ).

The point, why the use of Normal (Gaussian) distribution is wrong, is (again) hidden in "the context of the things". Similar to the example with the water on different

[20] As examples of the Power Law probability distributions, I can mention Pareto, Cauchy, Exponential distributions, etc.

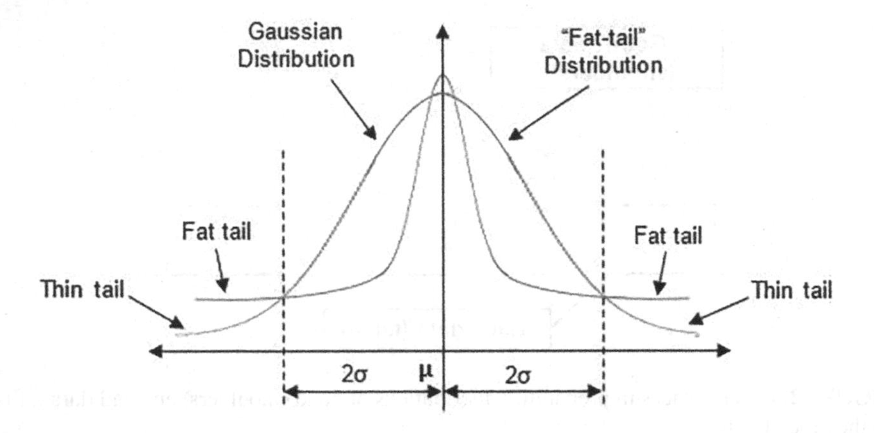

FIGURE 2.5 "Fat-tail" distribution compared to the Normal (Gaussian) distribution with the areas where "tails" start.

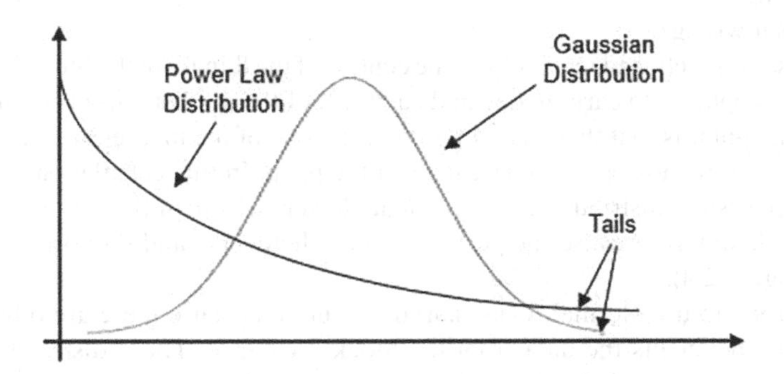

FIGURE 2.6 Power Law distribution compared to Normal (Gaussian) distribution.

temperatures in Figure 1.1, which distribution will be used to investigate (process) the data depends on the type of the data available. There are plenty of tests in statistics to determine which distribution can be associated to the data of interest. One of the important things is that most of the incidents and accidents data (also the Stock Exchange data) abide to the, so-called, Power Law distributions (Figure 2.6), with "fat tails" and this should not be neglected.

In the "Fat-tails" distributions, the adverse events are not treated as outliers. In fact, they can be found in the tails of distributions and, as such, they determine the shape of the curve. So, they should be used also in the Risky Industries because they are not treated as outliers in these distributions.

In general, in the Stock Exchange, the "normal" distribution is the one where the mean and standard deviation can be clearly determined and that is the reason that Normal (Gaussian) distribution is mostly used. The "Fat-tails" distributions are those where their standard deviations are infinite or cannot be determined (cannot be defined).

Determining the probability distribution is not an easy job, even for the Stock Exchange, nevertheless there is plenty of data. The problem is that there is a possibility of a few probability distributions to give positive answers for the statistical tests and you cannot know which one is OK. This and the fact that (maybe), the traders or employees in the Stock Exchange companies are lazy to calculate, with due attention, the probabilities, is a reason for NNT's disappointment.

Anyway, NNT is right: In the Stock Exchange area, "the context of the things" regarding available data and their distributions, is not used.

2.5 THE OUTLIERS AND THE BLACK SWANS EVENTS

Having in mind that the outliers can be treated as BSe in the statistical data and (by definition), all BSe are always outliers in the data, let's consider a little bit the situation with outliers…

The outliers, in the statistics, are defined as outcomes from the event (presented by data), which are considerably different from all other data in the particular distribution. As such, they do not belong to the data that describe the experiment, or, in other words, they belong to P_{NA}. Outliers may be caused by chance, wrongly chosen sample, error[21] in measurement, error in experiment, etc. As such, they need to be discarded.

But they also may point that we have wrong distribution in consideration: Maybe the outliers do not belong to assumed distribution (mostly Normal distribution), but they belong to "Fat-tails" distribution. This is a case when the wrong choice of the type of data distribution is made.

So, going back to the explanation in Section 2.2 (Can the BSe Be Considered in the Risky Industry?), especially as in Figure 2.1. Investigation of the BSe in the Risky Industries must be connected by the investigation of the outliers of data.

By the Black Swan definition of NNT, my understanding (having in mind the fact that all BSe are outliers[22]) is that the BSe must be found in the spaces which are far away from $\pm 3\sigma$ and there, for Normal (Gaussian) distribution, the probability of happening is only 0.27%.[23] Comparing the BSe (outliers) with the "Fat-tails" distributions (tails starting at $\pm 2\sigma$), in areas outside $\pm 3\sigma$, it can be noticed that there is considerable difference between the Normal (Gaussian) distribution and the "Fat-tails" distributions. Looking at Figure 2.5, this significant difference can be noticed even starting from $\pm 2\sigma$, and it is going to be bigger and bigger under $\pm 3\sigma$. In the scope of defining the BSe in the Stock Exchange, these events happen exactly in these areas where probability to happen is determined to be bigger than $\pm 3\sigma$ distance from the mean (μ). It means, NNT is right when he states that using the Normal (Gaussian) distribution in the cases where data abide to some other ("Fat-tail") distribution, the errors are considerably bigger.

---

[21] Reason for the error can be methodic, human, systematic, random, etc.

[22] Even other authors do not have problem with the NNT statement that BSe are outliers!

[23] To be realistically and scientifically correct, in the literature you can find definition of BSe as events (outliers!) which happen in area which is bigger than $\pm 6\sigma$. Probability to have such an event is actually 3.7 events in 1 million outcomes.

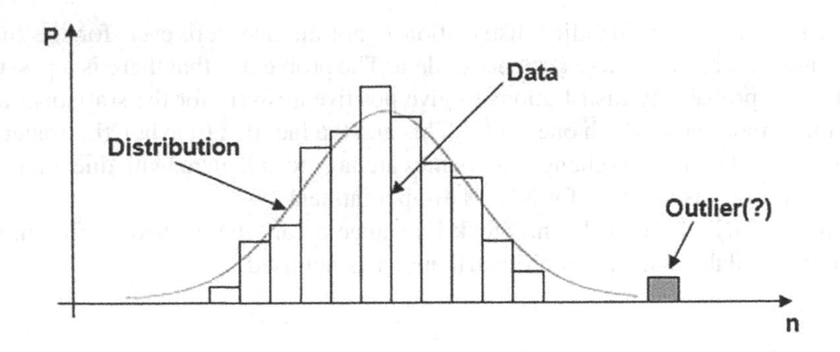

FIGURE 2.7 Data, distribution, and outlier.

But speaking about the statistical testing to check if some data is connected with particular distribution, the outliers are usually rejected for the test. This is not just simple rejection, but there are statistical tests to check if a particular outcome is an outlier or not. These tests give the probability that the particular outcome comes from another set of data. If this is the result of the test, the outlier can be discarded in this case. Anyway, the final decision of if the outcome is to be discarded or no must be done by the statistician.

So, the prudent statistician (investigating the particular sample of outcomes for particular event) should check first, if the possible outliers can fit or cannot fit in the data (Figure 2.7). He will use particular statistical tests for outliers, and he will check the outlier.

The problem here is that the statistician will work with sample of data. In the sample of data, the outliers can show up or they can be missed randomly (because choosing a sample is random process). So, whatever the prudent statistician does, the result will be uncertain with particular amount of probability. After checking and rejecting outliers, the analysis of data may continue with statistical tests if the data fits a particular probability distribution.

There is another important thing about outliers…

The outliers can belong to two sets of data in the same time, and it will make them very important, especially in the safety areas. Such a simple example is when we have binary decisions in digital electronics. Let's see the example from digital TTL circuits (Figure 2.8).

In TTL (Transistor-Transistor Logic) digital electronics, we send by the transmitter coded messages as a combination of two signals: "1" (as "high" voltage level) and "0" (as "low" voltage level). The criteria for decision (is it "1" or is it "0") are given on left side of Figure 2.8. So, if the voltage of the received signal is between 0 and 0.5 V, the received bit is "0". If the voltage is between 2.7 and 5 V, the bit is "1".

But, in the communication lines, the noise, the temperature and other factors affect the level of the signals and, on the receiver side, when the decision needs to be done, we have different levels from those which help us to decide is it "1" or "0". This area of uncertainty is pointed by the arrow on the diagram on the right side of Figure 2.8.

Looking at the right side of Figure 2.8, there are probability distributions for "0" and "1" and as it can be noticed, there is an (shaded) area of uncertainty ("0"

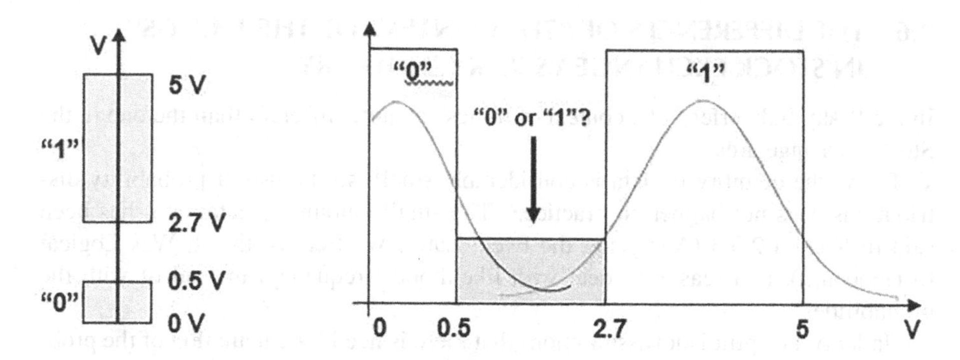

FIGURE 2.8 The TTL logic levels for decision (left) and associated probabilities (right).

or "1"?) where the decision cannot be made. In this area, where the outliers of both distributions are, it can be noticed that there is overlapping of the probability distributions. Simply put, in that area, we could not know if the outlier there belongs to the "0"-distribution or to "1"-distribution.[24] So, uncertainty here is big and, if in this area there is an outlier in the data, it must be removed. How the removal of the outliers from the available set of data can be done depends on the additional data embedded into it. This particular technique is used in digital telecommunications by adding additional bits determined by particular algorithm.[25] The concept of functioning of TTL circuits in communication is based actually on removing outliers (wrong messages) in the signal by the use of additional (redundant) information.

There is another important thing about outliers. Referring to the explanation of the probability in Section 2.3.1 (It Is Unpredictable…) in this book, it has been stated that:

> *"…in the real life even if we have "0" (0%) probability, the event may happen and even if we have "1" (100%) probability, the event may not happen."*

This statement can be also connected with the outliers. We cannot really know if the outlier is part of the data or it is just part of imperfection of probability. The uncertainty is still (and will be always) present, even in the cases where the prudent statisticians will do everything by the book.

In the Risky Industry, the outliers can be randomly present as it is in Nuclear Industry and Medicine, where radioactive materials are used. It is well known that the radiation is a random process and that there will be spikes of radiation which could have extreme levels.

---

[24] This is an issue in digital electronics, but it has been noticed very early and today there are excellent algorithms which can handle this problem of uncertainty!

[25] In digital telecommunications (each message is sent as coded stream of ones and zeros), there are plenty such techniques with redundancy. One of the first and simplest is so called "parity". Message sent by transmitter is divided in bytes (8 bits = 1 byte) where 7 bits belong to the message and the last one is "parity" bit. When the number of 1s in the byte is even, the 8th bit is adjusted to 1. If the number of 1s in the byte is odd, the 8th bit is adjusted to 0. If there is discrepancy between the numbers of 1s and adjustment of the 8th bit, the receiver will request from transmitter to resend the message.

2.6 THE DIFFERENCES OF "THE CONTEXT OF THE THINGS" IN STOCK EXCHANGE VS RISKY INDUSTRY

In the Risky Industries, "the context of things" is quite different than the one in the Stock Exchange area…

There, the quantity of data is considerably small, so the use of probability distributions does not happen in practice.[26] The small amount of data, as it has been said in Section 2.3.3 (Analyzing the Event Later, We Realize that It Was Logical to Happen…), is a reason to deal with likelihood (frequency) instead of with the probabilities.

Under (very optimistic) assumptions that there is need for calculation of the probability in safety, things will not be so complex regarding the dynamics of data. In the Risky Industries, the risk assessment is mostly based on three factors:

a. Determination of frequency (how often adverse events can happen);
b. Determination of severity of consequences (will someone die, be hurt, or just equipment damage will happen); and
c. Criteria using these two things which will help to make a decision: Is the system (process, operation, activity, etc.) safe or not?

It is clear from above that, for one adverse event, few consequences may exist (during car crash, someone can die, someone can be hurt, car could be destroyed, car could have only small damages, nothing could happen to anyone, etc.).

The probabilities which can be calculated for above three factors are:

a. Probability of the bad things to happen;
b. Probability for each consequence (which one will show up) if some adverse event happens; and
c. Combined probability from a) and for each b) from the previous two probabilities.

Maybe, the third probability will result with probability distribution which is with "fat tails", but this is again subject to the calculations for a particular situation. I do not have any clue how it will look simply because I have not found any book or article dealing with that.

There are few other very important aspects of the BSe in Risky Industries, which are different when compared to the Stock Exchange. In this paragraph I will try to discuss these.

In the Risky Industries, there is particular synergy between the Humans and Equipment, which is provided through the Procedures in everyday operations. In the Stock Exchange area, the Equipment is not included in the operations. There, the operational subjects are only Humans and this is something that makes a big difference.

---

[26] In my hitherto career in the area of Safety Management System, I never heard that some of the safety guys have mentioned "probability distribution"!

The Stock Exchange is based on the human behavior, which means it is prone to unpredictability of the humans. They, very often, must bring some decision based on the data that cannot be sufficient to get good result. In addition, even if the data is enough and with good quality, the humans, intentionally or unintentionally, can decide to do something else. In other words, intentionally they will undertake particular risks with the expectation of bigger gain or unintentionally they will make mistake under the influence of some of the Human Factors (HF). In addition, the changes which will affect us may be subject to someone else's decision (our manager's decision for example). This situation is actually present, not only in the Stock Exchange area but also in all aspects of our lives. We simply cannot run away from this.

The changes in our private or professional lives are stresses for us. Humans are basically reluctant to change. Even, when the changes will bring benefit, still, the uncertainty from the result of the change will be a source of stress. That is the reason that, in the Risky Industries, the Humans have established principles and rules that need to be followed through Change Management activities. Implementation of some management systems in particular areas (quality, safety, environment, security, change, etc.) is done with intention to mitigate the decision-making process of the Humans.

But "change" is not the only problem. Also, the "status quo" is a problem. Naturalists would like to say that if there is no flow of water (rivers), the water will be transformed into swamps. The danger of "status-quo" in the industry is based on the premise that, if there is no dynamism in the company's activities, something is missing. At the beginning, people like the "status-quo", but if it last too long, people will become lazy, and the adverse events "like" such behavior. Whatever is the "status-quo" in the company, the employees (especially Safety Manager) must understand that "the devil does not dig holes; he is focused only to make our lives terrible".

Another thing which makes difference between the BSe in the Stock Exchange and the BSe in the Risky Industries is the fact that humans would (more likely) put efforts to the possibility of earn money than efforts to avoid a disaster. This is our cognitive bias and we cannot run away from it.

In the Risky Industries, the Humans are the same as the humans in Stock Exchange, but they differ by the type of the education. The education of Humans in the Risky Industry is mostly in the area of something which I call "scientific engineering". These are usually engineers of different profiles, and their area of activity is, pretty much, deterministic (physical laws). The physical laws applied there are strongly independent of the human behavior or political situations. The employees there do not bother themselves very much with statistics and probability during their daily jobs.

From another side, the education of traders in the Stock Exchange is also scientific, but their daily jobs depend strongly, not on the physical laws but on the human behavior or political situations. And the human behavior or the political situations are full with uncertainty.

In the Risky Industries, the employees are affected by two other factors: Procedures and Equipment (technical systems used to manufacture the product/service) that do not exist in the Stock Exchange with the same "context of the things". The employees in the Risky Industries are not very often urged to make particular decisions, because, in general, the products and the services offered are given in cooperation with the Equipment and by following the Procedures.

The decision-making process in the Risky Industries is subject of three possible sources of uncertainty (Humans, Equipment, and Procedures). The clearest situation is with the Equipment. The Equipment is most reliable part of these three sources of uncertainty. The Equipment abides to the physical laws and is extremely predictable. The Determinism there is very strong, and the Randomness is mostly absent and the uncertainty is very low.

"The context of the things" in the Stock Exchange also differs from the point of view of how often the bad things happen. Due to human interaction, the catastrophic crises (BSe) in the Stock Exchange are mostly periodic, with the period of happening from 8 to 10 years. In the Risky Industry, there is no periodicity in the bad things happenings. There, the bad things happen as a sequence (chain) of events or when there is considerable lack of abiding to the rules and/or procedures.

The difference in "the context of the things" can be emphasized also by the different reactions of the BSe in the Stock Exchange and the BSe in the Risky Industries. When a BSe happens in one Stock Exchange, it will spread to the most of the Stock Exchanges in the country and in the world. After that, usually, the things will continue on the same way and nothing will be changed. Most of the experts will provide their own explanation about the root-cause, but nobody will care about that. The next day, everything will be the same.

But in the Risky Industry, every accident will trigger Regulatory investigation, which cannot be "bribed" due to many factors involved in the investigation. The investigation will try to find the root-cause and it will provide (if needed) new regulations/rules or recommendations on how to change the old rules with the intention to stop those things from happening again.

One very significant difference in "the context of the things", is the fact that, in the Risky Industries, there are plenty of elimination and mitigation measures, called "defense systems". These are part of the management system with the intention to eliminate hazards and eliminate or mitigate the risks. These "defense systems" are actually guarding the company from the adverse events and they deal with the consequences, if the adverse events happen. These are mostly specific to the company and there is general regulation which require a company to implement them.

In the Risk Industry areas, these "defense systems" are mostly implemented through regulation, and the companies there are using all available opportunities to earn money taking care of these regulations. In the fight for profit, the companies would not limit their chances to be safer, by introducing measures that differ from that requested by the Regulator. Some would say that it is not true. Please have in mind that, in the Stock Exchange, the implementation of specific measures different than regulation is exception, but in the Risky Industries, this is a well-established habit. So, in the Risky Industries, "the context of the things" regarding the BSe is quite different then it is in the Stock Exchange areas and it has its own influence on the nature of the BSe there.

The real question is: Can the Risky Industries learn something about the BSe in the Stock Exchange, which can help them to deal with the BSe in their area? This book is an attempt to answer this question…

2.7 INFLUENCE OF THE COMPLEXITY IN THE RISKY INDUSTRIES

We must understand that the social activities of humans are, also, affected by the equipment. Technology is growing very fast, beyond our expectations, and the social development of the humans cannot catch the pace with the equipment development. This was already mentioned at the beginning of the book. The new technologies are requesting different levels of education of humans and, to be honest, this level of education and understanding of the new technology is a problem for most of the humans (unfortunately).

Another aspect of the uncertainty, built by the Equipment, is the one which I call "safety paradox". Having in mind that the Equipment is more reliable than the Humans, almost everywhere in the industry, the Equipment "replace" the Humans. The level of automation and introduction of computer-based processing is huge and the human's influence is not seen or felt in most industrial processes. Today, in this highly automated and computerized industry, the Humans are still used to make decisions based on data coming from monitoring and control devices in the processes. It means that the Humans can still introduce uncertainty even in the outcomes produced by the highly automatic and computerized Equipment.

Whatever you are thinking about computers, they are stupid machines. The computers can do almost anything, but only if they are equipped with particular programs. This program actually "commands" computers about what to do, and these commands are based on human input. There are plenty of mistakes made by Humans where Equipment made the damage just because it was "commanded" wrongly. In 2018 and 2019, there were two crashes of Boeing 737 MAX aircraft, and the cause of the crashes is a simple (and sad) example of that.

There is another aspect of this automation and computerization: The presence of the computer-based processing is increasing the level of complexity of the Equipment; so, when there is a problem with the Equipment, it is hard to deal with it. Simply, with highly complex Equipment, finding and rectifying the faults is also complex task.

From "the context of things" of Gray Rhinos events (GRe),[27] the complexity of the Equipment could be a reason, or at least, it could contribute, the GRe's harbingers not being recognized. In such a case, the GRe cannot be stopped, and it can show all its power.

There is one curiosity considering the complex systems: Processing of the data, regarding failures of complex systems, have shown that the failure will happen more often than success, when humans try to intervene at the complex systems. The important thing to remember is that these types of failures are not random because they are introduced by humans. Anyway, they are based on common patterns, such as lack of understanding on what is going on, insufficient preparation to deal with the system, failure to determine external effects, wrong interpretation of the system's response on the changes, and so on.

The complexity is also the core of the "safety paradox": Trying to deal with Human unreliability from safety point, we produce complex equipment, which is another safety issue. Trying to increase the safety in one area, we actually could decrease the safety issues in another area. In other words: Trying to solve some of the problems, we create new ones. The increased complexity provides additional uncertainty in the Equipment

[27] Will be explained later in the book!

area and affects the Human's decision-making process. This is very much evident in the fault diagnosis and fault fixing of the complex systems. Time taken to fix the faults and the failures in the complex systems is considerably bigger than dealing with those in simple systems. In addition, the requirement for better education of Humans dealing with high automation and computerized processes is considerable high.

The complexity is also not good for the safety due to possibility of Chaos. As it has already been said, it can produce totally unpredictable behavior of the system, and you cannot control the system anymore.

The complexity, from "the context of the things" in the Risky Industries, very often is attempted to be resolved by reductionism. This is an approach where the complex systems are described through their constituent elements. The issue with this reductionism in the safety area is that the complexity of the system is presented as a sum of their parts. Actually, this is elementarily wrong, because, as mentioned before, the complexity does not necessarily depend on the number of the parts but on the complex interactions between them, without accounting for their number.

Anyway, there is a rectification for increasing the complexity (and uncertainty) of the Equipment. The problems with increased safety issues in the area of Equipment is more predictable, they are not so big and they can be handled also easier than the previous problems caused by the Humans.

2.8 THE EQUIPMENT AND THE BSe IN THE RISKY INDUSTRIES

As mentioned in the previous paragraphs, the BSe happens mostly as a result of the imperfection of the humans, and that is the area which is actually covered by NNT book. But the book, which you read at this moment, is dedicated to safety in the Risky Industries and there, beside the Humans, the Equipment and the Procedures are also part of the system.

I am very much disappointed by the ignorance which is manifested in the industry regarding production of good procedures, but I will not speak here about the impact of badly written procedures to the happening of BSe. The Procedures are written by Humans, and the Human impact on the BSe can also include the production of bad Procedures.

Regarding the Equipment, you will say that it is also produced by the Humans. Why we do not put it in the "basket", together with the Humans?

The answer is simple: Although both are products of Humans, there is one very big difference. The Procedures which are produced by different persons, although they are written for the same process, could be different. This cannot happen with the Equipment, because the Equipment is built based on particular physical law. These physical (let's say: natural laws) are the same for each person. The Humans also interfere with the Equipment, but the physical laws are still the same. Maybe there are some differences in the details or used materials, but (again): The used laws to build Equipment are the same.

For example:

a. Whatever is the rocket to put the spaceship in the space, the laws which needs to be satisfied regarding the navigation are same for each manufacturer;

b. Whatever is the medicine from the pharmaceutical companies, the chemical process of producing a medicine abides by the same chemical laws;
c. Whatever is the number of the cylinders in the car engine, each cylinder is using the same laws for combustion of the fuel and the principle of moving the camshaft is the same;
d. Etc.

In conclusion: Whatever the equipment is, the manufacturing (production) process is based on deterministic laws. As I said before: There is nothing random with the Equipment. There will be always some amount of uncertainty, but the Randomness will not exist. The uncertainty can be increased only in the situations where dynamical states of the equipment could enter the area of Chaos. The industry knows that and, the people there, have produced methods to deal with this uncertainty. The most used characteristics of determining the uncertainty of equipment's functioning is the reliability.

The reliability (*R*) is defined as probability that, for particular period of time, the equipment will not experience any fault (will work without fault), if at all, at the period before beginning of this time, it was properly used and maintained and it was normally operational.

There is mathematical equation used to calculate this probability:

$$R = 100 \cdot e^{-\frac{t}{\text{MTBF}}} \ (\%)$$

where *t* is time for which we are calculating reliability; and **MTBF** is Mean Time Between Faults which can be calculated by this equation:

$$\text{MTBF} = \frac{\text{AOT}}{n}$$

where **AOT** is Actual Operational Time (duration time when equipment is used for normal operation) and *n* is number of faults during **AOT**.

In some literature, you can find the failure rate (λ) as a measure for reliability, and there is nothing wrong with that. Actually, the reciprocal value of MTBF is equal to λ:

$$\lambda = \frac{n}{\text{AOT}} = \frac{1}{\text{MTBF}}$$

These are simple equations, but the reliability calculation is not so easy. I have not seen a book regarding reliability less than 300 pages in length. The problem is calculation of MTBF of the equipment, because it must be done before the equipment is put into operation and it is done by manufacturer. Whatever parts are used to build the equipment, each of them has a particular reliability (a probability that it will operate correctly for particular period of time). Putting all these parts into the system (Equipment), their interaction will combine and form total reliability of the Equipment. There are few books (manuals) where the reliability of different mechanical, electrical, and other parts (used to build complex systems) are published, and there are particular computer applications that can help with all these calculations.

There are few documents that provide data for reliability calculations and they can be used for calculating reliability. You can find many of them on internet for free download. The most famous is MIL HDBK 338B, "Electronic Reliability Design Handbook"[28] issued by Department of Defense in USA. Few other very useful books for electronic equipment are Telcordia[29] SR 332 (Reliability Prediction Procedure for Electronic Equipment), MIL HDBK 217F (Reliability Prediction of Electronic Equipment), Failure Rate Data Bank (FARADA), and RADC Non-Electronic Reliability Notebook.

So, the calculation of reliability of the Equipment is theoretical when the design of the equipment is finished, but it must be monitored by the user of the Equipment when the Equipment is installed on the site. When the reliability becomes too low, it is time to replace the old Equipment with new one.

In the industry, when the Equipment is installed and used for operation, the reliability is not calculated by the equations above, but the measure of the status of the equipment (normal or fault) is calculated thorough MTBF. Normally: Bigger MTBF means bigger reliability.

But even here the things could be vague...

Not every industry and not even every company uses the same terms regarding reliability. Some are using MTBF with meaning Mean Time Between Failures, some are using MTBO with meaning Mean Time Between Outages, to name just a few. Calculating and stating the reliability is very much used as marketing from companies. I prefer to use MTBF as Mean Time Between Faults because I made difference between "fault" and "failure". "Fault" is when the Equipment fails and "failure" is when operation (process, activity, etc.) fails. Not necessary operation fails due to failure of equipment. If you give wrong commands to the computer that controls the process, the computer will execute correctly, but the process will fail. So, there is no fault, but there is failure of operation. The same will happen if the Equipment is wrongly adjusted.

As shown in the last equation in this paragraph, the MTBF is connected by failure rate (not by fault rate), which can mean using "failure" instead "fault". Whatever you do and whenever you use the reliability, be careful: There is difference in all these definitions. So, when you would like to investigate reliability of your equipment, clarify with the manufacturer how they define reliability and how they calculate it.

There is another thing which you need to be aware of...

Reliability is about probability; the manufacturing companies provide only information of the MTBF to customers. The reason is that maintenance of the system can affect the MTBF values. So, whatever the value is provided for any Equipment, usually the MTBF value is very optimistic.

---

[28] Guidance material only!

[29] The company called Bellcore changed his name in Telcordia, made a revision of MIL HDBK 217 for telecommunication purposes and has published it as SR 332.

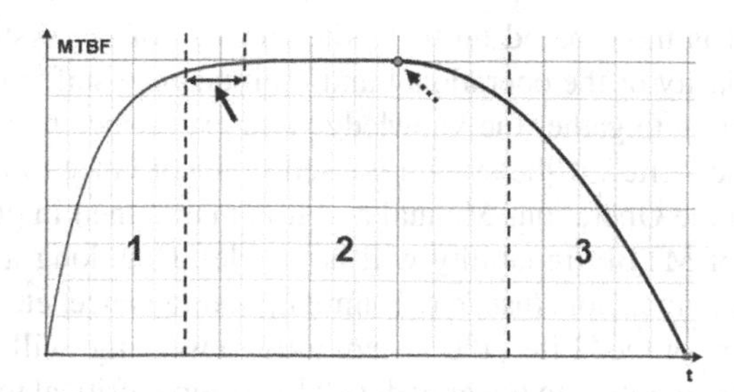

FIGURE 2.9 Change of MTBF with time (life cycle of system).

When Equipment is purchased and installed, the customers have a particular time to prove or disapprove this value. In the aviation, this period is 2 years. Anyway, it is not enough because it assumes that in these two years (in total 17,520 hours), the equipment with MTBF with 10,000 hours should fail only once or twice, which could never happen in reality. Realistic period to calculate true MTBF of equipment is (in my humble opinion) 5 years.

To show the meaning of MTBF, we can use the case when we would like to calculate reliability for the period of using the equipment equal to MTBF ($t = \text{MTBF}$). Using the equation from above we obtain:

$$R = 100 \cdot e^{-\frac{\text{MTBF}}{\text{MTBF}}} = R = 100 \cdot e^{-1} = 36.77\%$$

The result above means that the probability that the system will function properly for the period of time equal to the manufacturer's calculation of MTBF is 36.77%. Using ergodicity,[30] we can say that if we have installed 10,000 pieces of this type of system in our premises, after time $t = \text{MTBF}$, only 3677 pieces will function properly.

In Figure 2.9 is presented graphic of changes of MTBF (life cycle of system).

As it can be seen from Figure 2.9, the life cycle of single system through MTBF is shown and there are three areas.

Area 1 is area known as "Childhood," and it is area of design of the system and "improving of the design" process. Here, the MTBF is very low due to the design flaws that need to be fixed through testing. But as the design is improved, the MTBF increases in value. The time needed to deal with the product in the "Childhood" depends on the type of industry and complexity of the system. It can be in the range of a few months to a few years.

When the system is ready for production and selling, the dashed line presents start of area 2, which is actually process of selling and using of the system to the customer (installation). This area is known as "Life" of the system. It can be noticed (the full arrow) that, there is still MTBF which is not perfect (lower than established during the design) and there are still some faults. But these faults are not expression of the

[30] In statistic theory (used mostly for the purposes of economy), an ergodic set of data has the same statistical behavior averaged over time as averaged over the space. It means that statistical behavior of all set of data in one moment, is the same as statistical behavior of one element (of this set of data) over particular period of time.

bad system, but it is more based on the "adjustments" of the system "on the site" and on non-familiarity of the operational and maintenance staff to the new system. They need more time to gather the knowledge and experience how to deal with the system. During the "Life", if the system is used in accordance to the specifications and procedures in the Operations Manual and it is maintained in proper way, it will keep steady rate of MTBF (reliability will be stable). How long it will stay in this state depends on many factors (use, environment, maintenance, etc.).

In some moment of the "Life", the system (due to wearing) will start to fail more often (dashed arrow pointing to the gray dot). This is more critical to mechanical systems than to electrical systems. The electrical system (especially electronic systems) experience quite less wearing than the mechanical systems. The situation is still not too critical, but more monitoring and more maintenance will be needed. How much extra monitoring and maintenance will be needed, depends from the criticality of the operation. At some point, the system will be more in maintenance mode than in the operating mode; or the spare parts cannot be found anymore; or they are too expensive; so, it is cheaper for the system to be decommissioned than to keep it on site. This is area 3, called "Death" of the system. The company will need to purchase a new system which would be more effective, more efficient, and technologically more advanced. The buying of a new system not of the newest technology is, simply, not a wise move...

As it can be seen in this paragraph, due to the reliability, there is nothing in the Equipment which can satisfy "non-predictability" of the BSe from NNT definition. In any time of the life cycle of the Equipment, the good companies (in any industry) will have information how their Equipment would behave for particular period of time. What they will decide to do in regards to that information is something which has nothing to do with the BSe.

But there is also similarity with the reliability and the BSe: Both of them are described by "Fat-tail" probability distributions. By definition, the reliability of the Equipment is described by exponential distribution which is "Fat-tail" distribution. As time passes, the equipment is entering the "tails" of its life and this is a time where equipment should be replaced. If not, adverse events could happen.

As I mentioned before, the reliability is not easy to calculate. It applies to all industry, but my experience told me that approximately 75% of engineers are not aware that such a thing even exists. Nevertheless, there is no unique way how to do it, because it strongly depends on the configuration and the complexity of the Equipment. The art of using the reliability still stays with its calculation, but the most important is the interpretation of the results. This is the area where the behavior of the industry's managers resembles the behavior of the Stock Exchange traders: The profit is the most important even here!

Anyway, for the Risky Industries, the situation is quite different. There shall not be the same behavior of the managers regarding the profit in the Risky Industries and the Stock Exchange area simply because "the context of the things" is different.

Regarding the values of the MTBF in the Risky Industries, I can state, just as an example, the ILS (Instrument Landing System) in the aviation, where manufactures state that the MTBF is (theoretically calculated) not less than 14,000 hours. The ILS is a very critical navigational aid for landing the aircraft on the runways at the airport, but there are many others navigational aids that can stand less important that MTBF. In the aviation, to improve the reliability of the equipment, there is a rule that each critical piece of the Equipment, used for any phase of flight, must be duplicated. In addition, there must be duplicated monitors where each of them monitors performance of both pieces of Equipment, monitors the performance of other monitor, and monitors its own performance.

In general, duplication of the safety critical systems is highly recommended in every part of the Risky Industries. There, the reliability of the Equipment and the integrity of data are strictly and carefully monitored and controlled.

In the nuclear industry, I found some data on Internet and (using my simplified calculation), the MTBF of the nuclear reactor (encompassing all systems which help its performance) is approximately 17,500 hours.[31] Having in mind the impact of the catastrophic failure of the nuclear reactor to the humans and to the environment, it is understandable to have such a high value.

[31] This calculation was done by me using failure rate in OREDO data from document "Reliability Databases: State-of-the-Art and Perspectives" from Farit M. Akhmedjanov published in 2001 by Forskningscenter Risoe. Risoe-R, No. 1235(EN) in Denmark.

3 Analysis of the "Fat-tails" in the Risky Industries

ATTENTION: The Reader does not need to read this chapter to understand what is going on with the BSe in the Risky Industries!

For this paragraph to be understood, there is a need for good understanding of statistics and probability. Although I tried to simplify the explanations, I am not sure that I have succeeded in that. For me, the produced simplification seems to be alright, but it will not necessarily be OK also for you, as a Reader. Anyway, you can try. If you struggle to understand what is this about, just jump to Chapter 4. It is your choice what you will do!

3.1 INTRODUCTION

In the Stock Exchange area, most of the probability distributions of data are with "fat-tails". These are distributions characterized by big kurtosis and particular skewness. The high kurtosis contributes to the big gains (profit) and the skewness contributes to increase of the tails where the BSe are hidden. The kurtosis was explained in Section 1.5.3 (Randomness Vs Uncertainty) and in Figure 3.1 below is shown "skewed" probability distribution. As it can be noticed from the figure, the "skewed" probability distribution can be obtained by "beveling" right or left the Normal (Gaussian) or any other similar symmetrical probability distribution. This could happen in the Stock Exchange when the data changes very much and very often, but in general it is caused by more "ups" than "downs". Here, it is mentioned only Normal (Gaussian) distribution as most used, but have in mind that the skewness applies to all symmetrical distributions.

To better explain it, let's assume that there is a bus and there are 20 of my friends. If I use these 20 "elements" of the "set" and I try to present, by statistical means, the wealth in the bus, the distribution will be Normal (Gaussian). There are few of my friends who are wealthier than me and there are few who are poorer than me. on average, it will be Normal (Gaussian) distribution.

But, if we add Mr. Warren Buffet to the bus, (unfortunately, he does not even know that I exist!), then the Normal (Gaussian) distribution will become highly skewed towards Mr. Buffet. It will happen simply because all the wealth of my friends multiplied even by 1 million will not be comparable with Mr. Buffet's wealth. It is obvious that the skewness will be produced by the "extreme" wealth of Mr. Buffet, compared to the wealth of me and my friends.

So, as it can be seen, the level of "beveling" is a measure for skewness and it produces "Fat-tails" distributions. And the "fat tails" are the places where the BSe are hidden. In addition to the "fat tails", the skewness will decrease the peak of the distribution (kurtosis) which, in practice, will result with smaller gains (smaller profit).

DOI: 10.1201/9781003230298-3

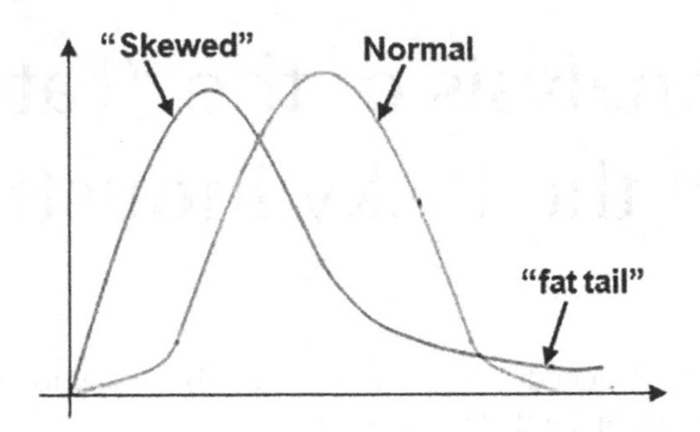

FIGURE 3.1 "Skewed" and Normal (Gaussian) probability distribution.

Doing research for the purpose of this book, I found the paper named "Black swans and VaR[1]" by Dr. Michael Adams and Dr. Barry Thornton (both from Jacksonville University) published in the *Journal of Finance and Accountancy*. In the abstract of the paper, there is something that can, highly, contribute to the different "context of the things" between the Stock Exchange and the Risky Industries in regards to "fat tails".

This paper speaks about two general classes of probability that are very different (qualitatively and quantitatively). The first class is based on Gaussian-Poisson distribution with "thin tail" and the second one is based on fractals, better known as Mandelbrotian[2] distribution (with "fat tails"). A Mandelbrotian distribution in the Stock Exchange is a probability distribution that provides high probability of experiencing small gains and small probability of experiencing a very large losses, which are very much bigger than the gains.

In the "Thin-tail" distributions, the hidden risks do not have large consequences. In the "Fat-tail" distributions, when the BSe happens, the consequences could be catastrophic.

In the text is presented significant characteristics of the Stock Exchange: The probability distributions of data, which are with "thin tails", usually do not produce big consequences, but probability distributions of data with "fat tails" usually have catastrophic consequences. As I have mentioned previously, the probability distributions explain how often some event will happen. With the Normal (Gaussian) distribution, most frequent events are in the middle, around the mean (μ) and less frequently in the tails.

In the Stock Exchange there is term "volatility" defined as change of particular set of data on daily basis (usually to worst). It is usually expressed through "changeable" standard deviation or variance.[3] In the cases where the data do not change extremely and not very often (volatility is low), the market behave normally

[1] VaR stands for Value at Risk. The paper can be downloaded from the link: https://www.aabri.com/manuscripts/131653.pdf (last time opened on 07/02/2021)

[2] Benoit B. Mandelbrot was a Polish-French-American mathematician and statistician working mostly in the practical sciences. Dealing with historical statistics of cotton prices and description of the sea and ocean shores, he recognized Chaos there.

[3] Squared standard deviation is called "variance" (variance = (standard deviation)^2 = σ^2).

and nothing significant will happen. This is because the normal fluctuations of the market during a day are nothing strange. The state of the Stock Exchange, in such a case, can be presented by "Thin-tail" probability distributions. But when the changes become big and more often, this is triggering panic with the traders and the cumulative (avalanche) effect will only amplify those changes for the worst. Then, the data probability distributions changes from "Thin-tail" into "Fat-tail" distributions. That is the situation with catastrophic consequences in the Stock Exchange area.

Of course, such things do not happen in the Risky Industries: "There, "the context of the things" is different.

This is the analysis made by me that I would like to present here regarding the "fat tails" dealing with the BSe in the Risky Industries. This is a considerable analysis which should explain aspects of the BSe and their connection to the "fat tails" for two cases.

The first case is connected with the influence of the protection measures which Safety Management provides to the systems into consideration (presented through Swiss Cheese model) and this is explained on Section 3.3. The second case is regarding the impact of different causes for the accident (BSe) and it is presented in Section 3.4.

3.2 SWISS CHEESE MODEL

In the Risky Industries, in the cases where the Swiss Cheese model[4] is used to explain the adverse events (BSe) in hindsight, the happening of accidents is defined as a "line of accident" (path) through the "holes in the barriers" in the safety systems (Figure 3.2).

Let's clarify something before I continue with the explanations...

Today, the industry is based on huge scientific and technological progress. The industry produces things and offering services which were unimaginable 50 years ago. The science and its associate, called "engineering", advanced so much that all these unimaginable products and services are complex by nature. So, to design, manufacture, implement, and maintain them, there are particular requirements for the employees involved in all these processes (operation, activities, etc.). These requirements come from the equipment which is used: It is really very complex by its structure and by its operation.

As I already have explained in Section 2.6 (The Differences of "the Context of the Things" in Stock Exchange Vs Risky Industry) and Section 2.7 (Influence of the Complexity in the Risky Industries), there are plenty of "defense systems" embedded in the management system and in the Equipment, which needs to improve the safety of the systems (processes, operations, activities, etc.) and the reliability of the Equipment. All these "defense system" make the Equipment more complex and this is something which I called "safety paradox".

[4] Although the Swiss Cheese's model is mostly used in hindsight to explain how the accident happened, it can be used also to calculate the probability of the accident to happen when barriers are installed. This model was proposed by Dr. James Reason in 1990.

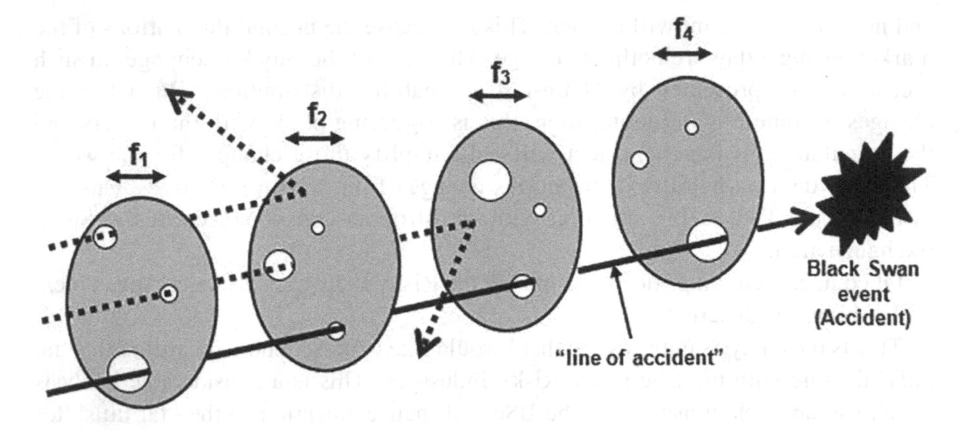

FIGURE 3.2 Swiss Cheese model.

The main point is that Swiss Cheese model does not take into account the system complexity, but it takes only the "defense systems" in place, which are treated as a system of barriers against the incidents and the accidents.

By implementing a Swiss Cheese model, we assume that each safety (defense) system is based on barriers which need to protect the system (operation, process, activity, etc.). The problem with all these barriers is that they are not perfect. There are always some "holes" in the barriers,[5] which are actually failures of the barriers to protect the system. The "holes" can be caused by the intrinsic imperfection of the barriers or by the changes of the states due to dynamic change of the barriers. Even when we establish the barriers, the "holes" cannot be precisely determined due to uncertainty of measurements and our assumptions, so we cannot know with high accuracy their exact position, their shape and how big is the "area" of the "holes". In addition, the barriers are dynamically influenced by internal (with other parameters in the system) or external (with the environment where the system is placed) interactions, so they will change in time and space and there is additional uncertainty regarding these changes of their positions.

If any of these "holes", can align in exact period of time or in the sequence of the time to the "line of accident", the adverse event can happen (Figure 3.2). It is assumed that it will happen in the same space, so the space will not be considered during the analysis.

The supporters of the Resilience Engineering (RE) assume that these barriers are compact by their structure, but their positions in time and space are dynamical (not stationary). They assume that the barriers oscillate randomly due to random (internal or external) interactions. These oscillations are presented as f_1, f_2, f_3 and f_4 on Figure 3.2. "Randomly oscillating" means that there is no constant frequency of oscillation, but there is one central frequency and there is a range of random changes around this central frequency.

[5] The "holes" in the barriers on Figure 3.1 are presented as circles, but their shape can be every possible shape!

Mathematically, as presented in Figure 3.2, the changeable frequency of oscillation can be presented as:

$$f_n = f_{on} \pm \Delta f_{\sigma n}$$

where $n = \{1, 2, 3, 4\}$; f_n is present frequency of particular barrier oscillation; f_{on} is central frequency of oscillation for the barrier n; and $\Delta f_{\sigma n}$ is a random change of the central frequency.

Having in mind that I speak about the random movement of the Swiss Cheese model's barriers, it is clear that each change of each frequency, will not abide to the same rule. If I try to connect these changes of the frequencies to statistical analysis, then f_n would be probabilistic expression for the dynamics of frequency for the barrier n, f_{on} will be average (mean) of all changes of that frequency and $\Delta f_{\sigma n}$ will be standard deviation of the change of that frequency. In the scope of this explanation, the equation above looks like equation for measurement uncertainty already mentioned in the Section 1.5.2 (Uncertainty):

$$R = \mu \pm 3\sigma$$

There are two possible cases regarding the "line of accident" alignment. The first one is connected to the alignment of "holes" which will align in the same time to produce the accident. The second case is regarding the alignment of the "holes" which do not happen in the same time, but the alignment is ordered as a "sequence of events[6]".

So, the "sequence of events" is when different holes align in quite ordered times to provide adverse event. I will explain the "sequence of events" with the example which I used in one of my first book (Quality-I is Safety-II: The integration of two management systems). This is a story which I read long time ago, but it is worth telling it here again.

When the "wise and old teacher" tried to teach his students about the need for order in their lives, he used a practical example. He brought one big glass jar and a sack of oranges and asked the students to fill the jar with the oranges, but taking care the number of the oranges in the jar to be maximal. When they finished, he asked them is the jar full and is it possible to put something else in the jar. The students said that the jar is full and it is not possible to put anything else inside.

The teacher took out, from under the table, a sack of rice. He poured the rice into a jar shaking it from time to time, with intention to maximize the quantity of poured rice inside. He asked the students again, is the jar full and is there any chance to put something else in the jar. The students answered: No!

The teacher reached the hand again under the table and took out a box full with salt. Students just looked how the teacher poured the salt inside, again shaking the jar during pouring. After that, he asked the students one more time: Is there possibility to pour something else inside the jar? The students could not find answer to this question guided by the previous experiences of the example. The teacher put his hand

[6] It could be also named as "sequence of states" assuming that particular ordered change of the states in time must happen to have an adverse event. Whatever you are using is OK.

again under the table and took out a kettle with hot coffee. He started to pour the hot coffee inside the jar. The hot coffee melted the salt and the dissolved salt provided space for more coffee.

It is obvious that if the teacher did not follow the particular "sequence of events" all these things could not be put into the jar. So, it is the same with the Swiss Cheese model: There is need of "sequence of events" to be aligned in time for adverse event to happened.

To be honest, the Swiss Cheese model is not so good a tool for "sequence of events" alignments, but having in mind that this is one of the ways the adverse event to happen, I will use it to explain it. The reason for that is the term "ergodicity" in statistics explained in Section 2.8 (The Equipment and the BSe in the Risky Industries). Applying ergodicity in the case of "sequence of events" will mean that we use the data for the events which happen in the same time and in the same space. And using the same data in time (ergodicity) corresponds to "sequence of events".

The "Sequence of events" (by my humble opinion) is the chain of events that most often happens as cause of the accidents and there is clear explanation why this happen. In the Risky Industries, there are clear procedures which cover all processes (operation, activities, etc.) and the employees there need to be trained in each of these procedures. The training is requested because there is a need of achieving particular "human automation" during execution of all industrial processes.

Going back to our Swiss Cheese model, the changes of the frequencies and the appropriate alignment of any "holes" with the "line of accidents", for both cases, must be in accordance with some probability distribution which is actually not Joint Probability Distribution (JPD), but I would call it a Combined Probability Distribution (CPD).

Theoretically there will be few probability distributions of our interest connected with the Swiss Cheese model presented in Figure 3.2:

a. Four distributions for random change of each frequency for each barrier (one per barrier). These distributions will actually correspond to the change of frequency of movement of each "hole" in the same barrier. We need these probability distributions to analyze the possible alignments of "holes" into the "line of accident";
b. Few distributions for each possible combination of these four distributions and for each of the few "lines of accidents" (which may exist). In Figure 3.2, there is only one "line of accident", but not necessary, the "holes" may align in more than one "line of accident". So, in general, there could be a few "lines of accident" (in the same time or in the "sequence of events"). In other words: It means that each of the few "lines of accident" will have its own probability distribution; and
c. One combined probability distribution for one set of events describing the probability for accidents when random alignment (combinations in space and time for all barriers and all "holes"!) of different "holes" and different "line of accident" happen. This will be actually the total combined probability distribution (CPD) of all possible "line of accident" distributions mentioned under (b).

Let's repeat again for better clarification: The situation (c) is depicting the alignment of different "holes" with different "lines of accident", simply because, more than one "line of accident", could exist in any system of barriers. In other words: The (c) is distribution of the set of events which are adverse events associated with all "lines of accidents" (all possible ways adverse event to happen). Analyzing this probability distribution means analyzing the adverse events in the scope of set of all possible adverse events which may happen in a particular system under consideration.

For our purposes, the first two situations (a) and (b), may be used to determine the "success" and "failures" of the operation that is guarded by these four barriers from Figure 3.2. This is because any situation when the "holes" are not aligned means "success" for our operation and any situation when the "holes" are aligned means "failure" (adverse event) in our situation. Of course, that the ratio between "success" and "failure" in our system depends on the shape of the area and number of the "holes": If they are big and many – there will be more "failures" and less "successes". In real life, we strive to make the area and numbers of each "hole" to be significantly small and with deterministic shape,[7] so the "failures" do not happen very often.

In Combined Probability Distribution (CPD), covering all these situations ("successes" and "failures"), the adverse events will be shown as outliers (they do not happen very often).

This is pretty much a very complicated situation because we need to find probability distribution for each dynamically changeable frequency and also for different combinations of different "holes" in any of the barriers, which also changes. The good thing is that it resembles very much the calculating uncertainties of any particular measurement. In metrology, for each measurement, we have different types of uncertainties and all of them may have different distribution. Eventually, when we have determined all distributions and all uncertainties, we can calculate the uncertainty for that particular measurement.

The point is that we cannot analyze c) because the accidents do not happen very often, so we do not have enough data, but we can analyze the movement (oscillations) of the barriers a) and their combination for different "lines of accident" b) and these analyses will produce a probability distribution which could be with "fat tails" and it will be also with big uncertainty.

Let's try to go into details in these situations…

3.3 SWISS CHEESE MODEL AND ALIGNMENT OF THE "HOLES"

This paragraph, to someone with considerable knowledge of statistics and probability, may look very strange and I am ready to accept his comments. The intention with this paragraph is to provide additional considerations of relations between the BSe and their happenings, so I would like to provide some examples which will bear the analysis. The chance that it would make someone not happy, is highly present…

[7] Having deterministic shape of the "hole" provide us with more control and less uncertainty.

I will explain few examples about the types of adverse events through explanation of their probability distributions, but (as mentioned in the previous paragraph!) I would like to use term "Combined Probability Distribution" (CPD) instead the term "Joint Probability distributions" which is associated to the systems[8] depending on two or more random variables which change constantly. In the real life, we try to control these variables and by controlling them, we control our process. The Joint Probability Distributions (JPD) are explaining how two or more random variables (parameters in the process) in the same time contribute to the process. The individual contribution to any of these random variables can be determined, but for our process, it is important to deal with CPD. So, each individual probability distribution[9] (IPD) can be expressed (mathematically) by two coordinates: Random variable (x) and probability distribution function ($f(x)$).

For a process with one variable, the probability distribution function will be expressed by a 2-dimensional system and it will be a surface in this 2-dimensional system. For a process with two variables,[10] the probability distribution function will be expressed by a 3-dimensional system and it will be a volume in this 3-dimensional system.

To provide just a simple example, I will use barriers from Figure 3.2 to express the contribution of their IPD in the protection of our system. The point here is that the functioning of each barrier may depend on one or few random variables, which, not necessarily, may be same for each barrier. My assumptions will be that there is only one variable for each barrier and they are different. In other words, the functioning of each barrier is independent of each other.

There is another thing in this paragraph which is not aligned with science. For the sake of science, I am speaking about adverse events that are discrete events, but I will use continuous curves to depict the adverse events. Again, forgive me, but this is only because, in my humble opinion, it is easy to explain "graphically continuous" events and to make you understand.

3.3.1 Alignment of the "Holes" in Same Time

The simple example of alignment of the "holes" in the barriers in the same time, is the mid-air collision over Uberlingen (Switzerland) between two flying aircraft. The collision happened on 1st of July 2002 between Ukraine's Tu-154 passenger jet and DHL's Boeing 757 cargo jet. There were few contributory events why this accident happened, but the main point is that, all these events happened in the same time and in the same space. In other words: The "holes" aligned themselves in the same time and in the same space.

Let's not go into details about this case but try to analyze alignment of the "holes" (events) in the same time.

---

[8] In general, I am using here term "system", but whatever is said in the book (with maybe small number of exceptions) applies also to operation, activity, process, etc.

[9] Somewhere in the literature, you can find it under the name "marginal probability distributions".

[10] For the processes with more than 3 variables, graphical explanation is not possible, but tables are used for explanations.

For this case, the total probability for accident (as number describing the probability, not as probability distribution!) will be the product of individual numbers for probabilities[11]:

$$p_{\text{tot}} = p_{hf1} \cdot p_{hf2} \cdot p_{hf3} \cdot p_{hf4}$$

where p_{tot} is total probability that all "holes" will be aligned in one "line of accident", and p_{hf1}, p_{hf2}, p_{hf3} and p_{hf4} are probabilities of each "hole" in barriers from 3.2, which will align themselves in the same "line of accident".

The main point here is that, nevertheless, we have multiplication of probabilities there, the total probability will decrease. The reason is that the probabilities calculated are between 0 and 1 and as such, their multiplication will result in total probability which will be considerably smaller. You may not be surprised because this is something which is well known in the Safety field. The experience (and calculations) showed: "More barriers = more protection". Of course, here we do not need to exaggerate with the number of barriers, because each barrier is additional complication in our system and as such, it will create additional hazards. Additionally, in the case of failure, the fault-fixing and recovery of these systems with more barriers (due to their complexity) will be longer and it will need more resources ("safety paradox").

Let's calculate the total uncertainty regarding the positions of the "holes" in the barriers for the values of σ_n ($n = 1, 2, 3, 4$) for the barriers in Figure 3.2 assuming some values for each σ. I assumed: $\sigma_1 = 0.3$, $\sigma_2 = 0.2$, $\sigma_3 = 0.1$ and $\sigma_4 = 0.4$. All different $\boldsymbol{\sigma}$ should add on to each other, but as variances (previously squared):

$$\sigma_{\text{tot}} = \sqrt{\sigma_1^2 + \sigma_2^2 + \sigma_3^2 + \sigma_4^2} = \sqrt{0.3^2 + 0.2^2 + 0.1^2 + 0.4^2} = 0.54$$

As it can be seen from the equation, the total standard deviation (uncertainty) σ_{tot} will increase, but only for small value compared to the values of contributing standard deviations. In other words: The total value of uncertainty expressed by standard deviation (σ_{tot}) is bigger than any individual uncertainty, but it is bigger only 35% compared to the biggest individual uncertainty.

Looking the decreasing of total probability of multiplied probabilities, we can notice that it is considerably bigger. Under assumptions that all four probabilities are 0.1, the total probability will be 0.0001 which is decreasing of 1000 times.

In general, by putting the barriers, we will considerably decrease the probability that adverse event will happen, but the uncertainty of our solution will increase. And this is something which can contribute to the BSe: With increasing the number of barriers, we increase also the uncertainty, so the prediction of BSe will decrease.

Let's see what will happen if we are dealing with probability distributions instead of probabilities (as number) and uncertainty.

[11] Here I am speaking about discrete events with particular discrete probability distributions where each outcome (event!) has its own probability expressed as number. In the case of continual probability functions, there are no events and no individual probabilities, just interval probabilities are present with continual probability distributions.

For the sake of simplicity, I will deal only with probability distributions of only two barriers and they will be used to calculate CPD for the alignment of the "holes". This is not simple as the equation from above (for probabilities as numbers), because, in this case, the probability distributions will convolute. The convolution of two functions, $f(x)$ and $g(x)$, mathematically can be expressed by the following equation:

$$f * g = \int_{-\infty}^{\infty} f(x) \cdot g(t - x) dx$$

The probability distribution functions, $f(x)$ and $g(x)$ will be defined by the Gaussian equations:

$$f(x) = \frac{1}{\sigma_1 \sqrt{2\pi}} \cdot e^{-\frac{(x-\mu_1)^2}{2\sigma_1^2}}$$

$$g(x) = \frac{1}{\sigma_2 \sqrt{2\pi}} \cdot e^{-\frac{(x-\mu_2)^2}{2\sigma_2^2}}$$

where μ is mean of the set of data used to calculate the distributions and σ is the standard deviation of the data used to calculate distributions. In accordance with Section 1.5.2 (Uncertainty), μ is measure for accuracy (more data – more accurate assumptions) and σ is the measure of precision (uncertainty); bigger σ – bigger uncertainty!

The overall situation is graphically presented in Figure 3.3.[12]

As you can notice, in Figure 3.3, the combined (convoluted) distribution is smaller than other particular distributions of each of the barriers.

So, alignment of the "holes" in the same time, as it can be seen, will produce decreased probability and in our case (using the Normal (Gaussian) distributions), the convolution distribution will also resemble to Normal (Gaussian) distribution. I said "resemble", because it cannot be really Normal (Gaussian) distribution. For the sake of simplicity, I have chosen the values of μ and σ for both barrier's probability distributions, to make another point where this resemblance could provide additional information.

Looking the situation in Figure 3.3, we can notice few things:

a. Convolution distribution is smaller (lower kurtosis). If I use more than two barriers, it will be even smaller. This is expected result;
b. The convolution distribution is wider, which comes from the fact that the standard deviation (uncertainty) is bigger. Using more barriers will slightly increase the standard deviation. This is also the expected result;
c. Look at the tails: They are fatter compared with the tails of individual distributions! If we use more than two barriers, they will become even fatter.

[12] Please note that this is highly hypothetical and hugely simplified case! In reality things are pretty much complex and I am not sure that this simple analysis could apply there...

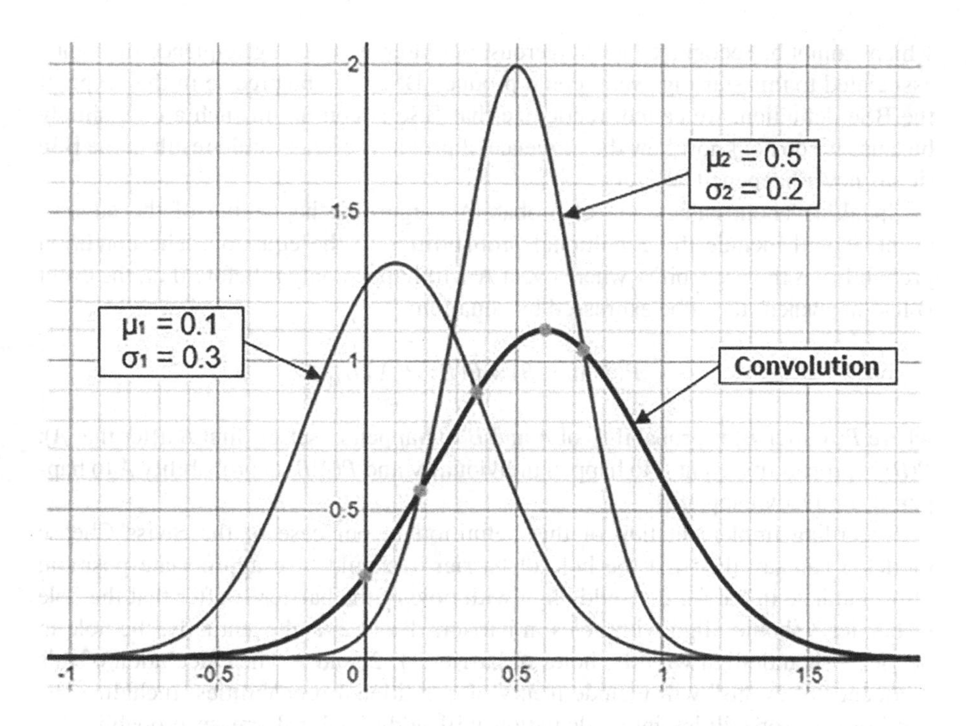

FIGURE 3.3 Probability distribution for two combined distributions (Convolution!).

So, the combination of few probability distributions will provide combined distribution which is with "fat tails" and this is the reason that, roughly, in the Stock Exchange area most of the distributions of data must be considered by "fat tails". There are many variables there and plenty of them are combined, when there is need to make a decision. In addition to the "fat tails", the uncertainty there will be also bigger.

3.3.2 Alignment of the "Holes" in Sequence of Time

As I mentioned before, the alignment of the "holes" could also happen if there is "sequence of events" (not in the same time). It means that overall cause will be: The adverse events happen as result of causes which, in particular ordered period of time, will align. It could be a good case, because it means that we can react immediately, if we notice some of the ordered events on time to "break" the sequence. This breaking the "sequence of events" on time could protect the system from a BSe.

A simple example for this case is the world financial crisis in 2008. It was started in 2006 by the sudden fall of the house prices in USA. It caused a sequence of events which could not be stopped because no one understood what was going on and it exploded in 2008, first in the USA and in then the rest of the World.

This case covers most of the events in the Stock Exchange, but it applies also to the accident in the Risky Industries. The BSe happen in the Stock Exchange due to "alignment" in sequence of many other events. It could start with some activity which cannot be recognized as dangerous, but later, an ongoing sequence of events, associated to this starting event, could produce BSe.[13] To be precise in the scope of the BSe definition, we cannot recognize that BSe is coming in such a case simply because we do not know how the sequence of previous events would result in the BSe (it has never happened before).

Speaking about calculation of probability regarding happening of the adverse event, it will include the conditional probabilities in the equation. The condition probability is the probability where event A will happen only if before that, the event B had happened. It can be expressed by equation:

$$P(A \text{ after } B) = P(B) \cdot P(A \mid B)$$

where $P(A \text{ after } B)$ is probability of A and B to happen in series (first B after that A); $P(B)$ is probability event B to happen individually and $P(A|B)$ is probability A to happen if B already happened.

Regarding implementation of this definition in our case of the Swiss Cheese model, I can say that first the hole of barrier 1 should take appropriate position; after that hole in barrier 2 should align with hole of the barrier 1; after that the hole of barrier 3 should align with holes in barriers 1 and 2; and eventually, the hole in barrier 4 should align with the holes of barriers 1, 2, and 3. The calculation of this "sequence of events" will include plenty of conditional probabilities. Including the conditional probabilities into calculation will additionally decrease probability of having an adverse event.

3.4 MULTIPLE CAUSES OF ACCIDENT TO HAPPEN...?

There is more than one cause for the accidents to happen, and there is more than one cause for the systems (processes, activities, operations, etc.) to fail. The car will not move if there is no fuel; if there is no battery (electricity); if the engine is damaged; if the distribution of the fuel of electricity in the car is damaged; etc. All these events will be in the tails of the overall probability distribution for the operation of the car.

In such cases, the Swiss Cheese model cannot be used for combined probability because it applies only to individual cases (one "line of accident"). I cannot produce combined Swiss Cheese model simply because in the case of different causes, I am speaking about different barriers that do not necessarily align themselves only in one "line of accident". It is understandable that there could be a few "lines of accident" and as such, any of them could have different triggering events (causes). The better way to analyze all the ways of how the systems (processes, activities, operations, etc.) would fail is to use Fault Tree Analysis[14] (FTA).

Let's see what will happen with probabilities in the areas of complex systems.

---

[13] This situation with "sequence of events" is more appropriate for Gray Rhino events (GRe) which will be explained later in the book.

[14] By my humble opinion, the best methodology for risk analysis is the Bowtie Methodology which consists of two methods: Fault Tree Analysis (FTA) and Event Tree Analysis (ETA).

The complex systems, for the purpose of this book, will be defined as systems which consists from many other subsystems (parts) and all these subsystems, working together, will provide a normal operation of our complex system. It means that failure or fault of any of these subsystems will cause a failure or fault of our complex system also. These failures or faults of the subsystems that cause the complex system to stop to provide the normal operation, I will call "critical failures" ("critical faults").

If we consider all probabilities (as numbers) regarding the all-possible situations (events) for the critical failure of any of the subsystems, the probabilities will add[15]:

$$p_{\text{tot}} = p_1 + p_2 + p_3 + \ldots + p_k = 1$$

where p_{tot} is total probability that the complex system will fail and p_k ($k = 1, 2, 3, \ldots, n$) are individual probabilities that any of the n-subsystems will fail. It means that calculating the total probability, I need to add individual probabilities that any of the subsystems will fail. The number one (1), at the end of the equation, means that, these are all possible situations (events, causes) of the failures of the operation of the system.

But speaking about probability distributions, you have to understand that we are dealing with functions. There, the probability (as number) for some particular event to happen is calculated based on the integration of the probability distribution for continuous functions or by reading or adding probabilities for discrete functions. It is explained in the Section 2.3.1.3 (Probability Distributions).

For this particular case of multiple causes for accident, there will be addition of the individual probability distributions (IPD) for describing the normal operation together with all possible cases of the failures of the system. In this situation, I would combine the probability distribution functions not for the barriers but for each operation of the individual subsystems presented by probability distributions for each subsystem. We have few probability functions (one per operation of subsystem!) and each of them can be combined into system by adding these probability distributions. How the summation (addition) of the IPDs will be done, I will not explain here because it is simply too complex and that is not the point.

Anyway, the CPD will describe the total operation of the system and it is graphically presented in Figure 3.4. Under the simplified assumption that our system is presented by four subsystems, I have used four individual probability distributions, each of them presented by different Normal (Gaussian) distribution with different μ and σ.

As it can be seen in Figure 3.4, the CPD (bolded line) is skewed on the right side and, as it has been already explained in Section 3.1 (Introduction), this skewness produces "fat tails". Looking at Figure 3.4, the normal operation of the complex system will be at the middle of the "combined" probability distribution and the BSe will be in the tails (they will be outliers).

---

[15] This is again subject of connection between the subsystems which are constituents of the complex system. The case considered in this paragraph is simplest one! The real picture regarding the causes of the failures of the complex system can be obtained if I use complete Fault Tree Analysis.

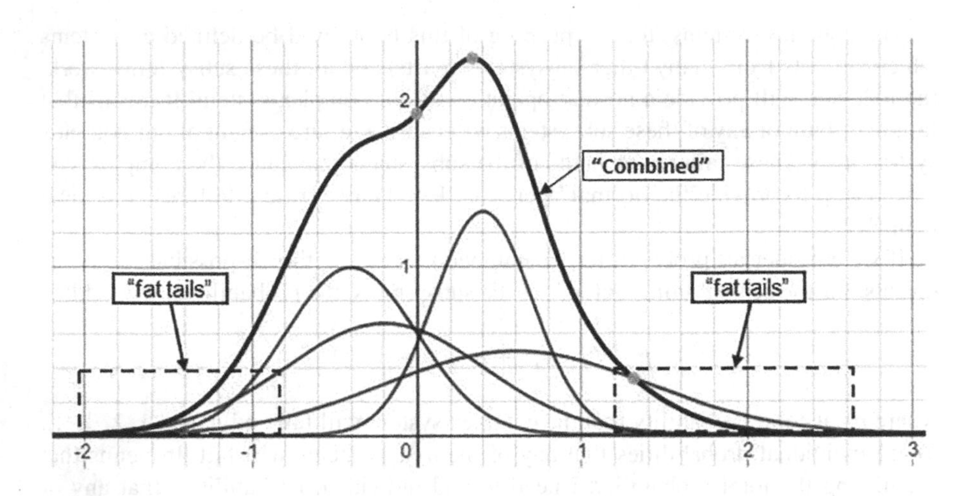

FIGURE 3.4 "Fat tails" of "Combined" probability distribution.

For this case, the uncertainty will be same as previous case because I will use the same equation for calculating total standard deviation (σ_{tot}) and four individual standard deviations, as it is done in previous paragraphs. In general, uncertainty will be bigger. How much is not important in this case.

That, what is very interesting here, is the fact that CPD is a "Fat-tails" distribution, although we were analyzing normal operation (presented by four of Normal (Gaussian) distributions). This is pretty much in accordance to the NNT statements, that the BSe are hidden in the "Fat-tails" distributions and using the Normal (Gaussian) distribution is wrong because the bunch of different events will provide CPD which cannot be Normal. So, NNT is right: The Normal (Gaussian) probability distribution applies only to simple systems (processes, operation, activities, etc.) and even in that case, not to each of them.

The main point of this paragraph is: Combining many events in the complex systems, which can contribute to the overall system operation, could result in the "Fat-tails" probability distribution of the behavior of the system. So, approximating the "Fat-tail" distribution of the complex system with the Normal (Gaussian) probability distribution could mask the increased risks in the tails, and this is the place where the BSe are!

3.5 MAY WE USE THIS APPROACH IN THE RISKY INDUSTRY?

This is a chapter where I tried to present a strange mixture of the situation in the Stock Exchange and in the Risky Industries. For those who are good in statistics, probabilities and safety, they can comment that this chapter looks pretty much unnatural...

I can agree, but again: This is due to the different "context of the things".

The Swiss Cheese model is used in Safety to explain that there is a need of "chain of events" (alignment of the "holes") adverse events to happen. This is not necessary in the cases in the Stock Exchange. Anyway, the calculations to undertake the risk in the Stock Exchange are based on the probability of distributions of abundance of data (which are highly dynamic). There, the primary mistake could be in the "modeling the data" or by using a statistician language: By making the wrong decision about probability distribution that does not correspond to the data under consideration.

To be honest, the "tragedy" of this wrong decision is more connected with the disappointment about expectations than by the decision itself. When the traders in the Stock Exchange area, undertake the decision to accept the risk, their expectation that the things could go wrong are small, because they are not based on "Fat-tail" distribution. So, any measure to deal with the risk, if they had made a wrong decision, is proportional to the risk which is underestimated. If the things do not go as expected, these measures will fail, simply because the expectation, based on the wrongly estimated risk, was wrong. This is actually the BSe: The event was not expected (due to wrong risk calculation) and it is catastrophic (due to the wrong measures to contain, eliminate or mitigate it).

"The context of the things" in the Risky Industries is different. Nevertheless, I tried in this chapter to give analysis of the probabilities and their combination, the things with the calculating of the risks are different there. In the Risky Industries, the risk is presented through the determined and quantified hazards: For each hazard, we calculate the frequency and severity, and this is the risk for that hazard. Many hazards – many risks!

So, the mistakes, which could be happen in the areas of expectations in the Risky Industries, are connected with the wrong assumptions regarding determined frequency and severity. Accepting the Stock Exchange's approach, for each adverse event in the Risky Industry, we could produce two probability distributions: One for adverse events in regards their frequencies and one for adverse events in regards the severity of their consequences.

In addition, maybe there will be need to provide additional probability distributions which would be convolution of both probability distributions (for frequency and for severity) because this is how the risk is defined in the Risky Industry.

But the point is that this approach does not work in the Risky Industries due to lack of data. The lack of data could not help, also, with the outliers. Even if we try to do it, the determination of type of probability distributions would not be determined due to this lack of data. We will not know is it "Fat-tail" or something else. In the cases when you cannot determine the probability distribution of the data, it is a "rule of the thumb" to assume that this is uniform probability distribution. In such a case, the uncertainty is highest because the probability of each event to happen is the same for every event.

So, whatever calculation we do, the uncertainty will produce inconclusiveness.

In the Risky Industries, the primary job is to produce product (offer services), and it must be done by taking care for safety of these products/services in regards humans, assets, and environment. That is the reason that, due to all safety measures implemented there, the adverse events do not happen very often and the results of processing the available data could be inconclusive.

4 What to Do with BSe in the Risky Industry?

4.1 INTRODUCTION

In the previous two chapters, considerable mathematical analysis and comparison of BSe in the Stock Exchange and in the Risky Industry was presented. I have presented plenty of data and situations which showed that "the context of the things" regarding the BSe in the Stock Exchange (NNT definition of BSe) and in the Risky industry cannot be the same.

This chapter presents the matrix used in the civil aviation field to calculate the risk. Similar approaches can be found in other Risky Industries.

Anyway, having in mind that the BSe in the Risky Industry can be associated to the incidents and accidents, we should take care to avoid them. This is not an easy job, having in mind that for decades, humans have been striving to stop the accidents but they still happen...

Let's see how we can maintain happenings of the BSe in the Risky Industry...

4.2 HOW WE CALCULATE THE RISK IN THE RISKY INDUSTRIES?

There are two types of safety which apply to every industry. The first one is Occupational Health and Safety (OHS) and the second one is Functional Safety (in the Risky Industries, known simple as Safety). Let's explain both with one simple example...

Imagine that I have a factory which produces cars. There is a building with a production hall and there are offices; there are machines, employees, and warehouses. All those things I will call "assets" in this chapter.

Whatever is inside the factory (assets) shall be safe for the employees, the buildings, the machines, the environment, etc. Whatever I do during the production process of the cars, it shall not produce any harm, damage, or death to any of the assets. The Regulatory rules that require the provision of all these things are known as Occupational Health and Safety (OHS) regulations. Very often, this type of regulation is combined with the protection of the environment outside of the factory, so it is also known as Occupational Health, Safety and Environment (OHSE).

Speaking about OHS means that I speak about controlled environment:

a. It does not matter how the temperature outside the factory is. I will maintain the temperature inside between 22°C and 25°C. If my factory is in Sweden, I will heat my premises during the winter, and if my factory is in Saudi Arabia, I will cool my premises during the summer;

DOI: 10.1201/9781003230298-4

b. It does not matter where the production halls and offices are and how they look. I must provide at least 6 m^3 volume of space to any employee;
c. It does not matter if it is windy or rainy outside: I must provide a quiet and dry space for my employees, machines, materials, and finished products;
d. It does not matter where the production halls and offices are and how they look. I must provide particular supply of water, kitchen (food, coffee, tea...), and toilets in capacity determined by the number of my employees and the structure of my operations (processes);
e. It does not matter where the production halls and offices are and how they look. I must provide particular fire protection which is determined by the Regulatory rules;
f. Etc.

All these regulations are well known and they apply not only to my factory but also to all industries, hospitals, schools, banks, financial or educational institutions, etc. So, providing the OHS is done through well-determined regulations based on previous investigation to increase effectiveness, efficiency, and safety of the processes in our everyday private and professional lives.

When the car which is produced in my factory (the controlled environment!) is sold to any customer, then OHS does not applies any more. My car can be sold in Sweden (where during winter temperatures fall to −40°C) and in Saudi Arabia (where during summer temperatures go up to +50°C). My customers will be the persons with different social, educational, religious, and cultural status, and my car must satisfy the safety of all of these customers in their specific environment.

This is a "duty" of the Functional Safety. The Functional Safety shall provide safety for the customers in any conditions which are not regulated, and this safety differs very much from place to place and from customer to customer. That is the reason the Functional Safety applies mostly to the Risky Industries. There, the regulation is requiring implementation, maintenance, continual monitoring, and improvement of particular Safety Management System (SMS). This type of Safety does not apply to other industries (banks, insurance, schools, etc.).

Regarding the OHS, there is an ISO standard: ISO 45001:2018. It is named "Occupational Health and Safety Management Systems – Requirements with guidance to use". But there is no ISO standard dealing with Functional Safety due to reasons mentioned above in this paragraph.

In the Risky Industries, the SMS shall take care how to protect humans, assets, and environment during use of their products and/or providing their services. The approach for this Safety is proactive: In advance, the companies establish a List of Hazards which are important for their situation, and they calculate the frequency of happening (likelihood) and severity of consequences. This is actually transformation of the hazards into risks. Risks are calculated for each of the hazards on the list, and the companies strive to fix (eliminate and/or mitigate) the risks before they happen.

There are plenty of methods and methodologies used to deal with all these things, but no one is dealing with probabilities and probability distributions. There is a simple reason for that: There is lack of data. In the Risky Industries, the bad things do not happen often, so their investigation is quite different than those in the Stock Exchange area (where data is abundant).

The automotive industry is using Failure Mode and Effect Analysis (FMEA). There are requirements in the IATF 16949 standard (which is dedicated to automotive industry), and most of the companies are certified in accordance with this standard.

The nuclear industry is mostly based on improving the quality of the processes which needs to provide safety. There, Human Factors are not so much present, because most of the processes are done by equipment which is highly automatic and computerized. Humans are mostly involved in monitoring and maintenance.

In aviation industry,[1] there is categorization of the frequency of happening and severity of occurrences (Table 4.1) and there the Risk Assessment Matrix is used (Table 4.2). Criteria for safe operation are given in Table 4.3.

As it can be noticed, in aviation, the risks are only qualitative (no numbers and no probability distributions for calculations). There are no quantitative calculations because it is too uncertain to deal with something which happens very rarely.

If we would like to be scientifically correct, there is need to calculate the probability to each of the rectangles in Risk Assessment Matrix (Table 4.2). Having in mind

TABLE 4.1[a]
Likelihood and Severity in Aviation

Likelihood	Meaning	Value
Frequent	Likely to occur many times (has occurred frequently)	5
Occasional	Likely to occur sometimes (has occurred infrequently)	4
Remote	Unlikely to occur, but possible (has occurred rarely)	3
Improbable	Very unlikely to occur (not known to have occurred)	2
Extremely improbable	Almost inconceivable that the event will occur	1
Severity	**Meaning**	**Value**
Catastrophic	Equipment destroyed, multiple deaths	A
Hazardous	A large reduction in safety margins, physical distress or workload such that the operators cannot be relied upon to perform their tasks accurately or completely, serious injury, major equipment damage	B
Major	A significant reduction in safety margins, a reduction in the ability of the operators to cope with adverse operating conditions as a result of an increase in workload or as result of condition impairing their efficiency, serious incident, injury to persons	C
Minor	Nuisance, operating limitations, use of emergency procedures, minor incident	D
Negligible	Few consequences	E

[a] All tables in this paragraph are undertaken from ICAO document (Doc 9859; Safety Management Manual; Third Edition, 2013).

[1] There is similar categorization in other Risky Industries too!

TABLE 4.2
Risk Assessment Matrix

	Risk severity				
	Catastrophic	Hazardous	Major	Minor	Negligible
Risk probability	A	B	C	D	E
Frequent 5	5A	5B	5C	5D	5E
Occasional 4	4A	4B	4C	4D	4E
Remote 3	3A	3B	3C	3D	3E
Improbable 2	2A	2B	2C	2D	2E
Extremely improbable 1	1A	1B	1C	1D	1E

TABLE 4.3
Risk Tolerability Matrix with explanations

Tolerability description	Assessed risk index	Suggested criteria
	5A, 5B, 5C, 4A, 4B, 3A	Unacceptable under the existing circumstances
	5E, 5D, 4C, 4D, 4E, 3B, 3C, 3D, 2A, 2B, 2C, 1A	Acceptable based on risk mitigation. It may require management decision.
	3E, 2D, 2E, 1B, 1C, 1D, 1A	Acceptable

black	High (intolerable) risk - Immediate action required for treating or avoiding risk! Cease or cut back operation promptly if necessary! Perform priority risk mitigation to ensure that additional or enhanced preventive controls are put in place to bring down the risk index to the MEDIUM or LOW OR NO RISK range.
gray	Medium risk - Shall be treated immediately for risk mitigation. Schedule for performance of safety assessment and mitigation to bring down the risk index to the LOW OR NO RISK range if viable.
white	Low or no risk - Acceptable as it is. No further risk mitigation required.

that each event with particular combination of Likelihood and Severity from Risk Assessment Matrix will have values which are calculated for a particular company and its environment, it must be done by the company itself. It means that they need to employ mathematicians (statisticians), but for the time being, the Regulatory Bodies and the aviation subjects have opinion that it is not economically and practically sustainable. All the traders in the Stock Exchange are qualified statisticians, and this is one more difference regarding "the context of the things" in the Stock Exchange and in the Risky Industries.

Anyway, in the next paragraph, there are two figures which show that SMS in aviation is providing good results, even without consideration of the BSe and the "Fat-tail" distributions.

4.3 BSe IN THE RISKY INDUSTRIES

The title of this chapter is: What to do with the BSe in the Risky Industries?

To be honest, the answer of the question in the title of this paragraph is: Maybe there is need for nothing special! The BSe events do not apply to the Risky Industries in the same manner as they apply in the Stock Exchange. It does not mean that there are no solutions, but they need to be implemented through the SMS implementation.

Let's try to explain this more thoroughly…

As it has been mentioned before, the BSe can be treated as a synonym for accidents in the Risky Industries, but the main point is that the BSe (as defined by NNT) cannot endanger the Risky Industries in any different way as it can do the accidents.

This is a very important statement regarding the BSe in the Stock Exchange area: They are quite different than the BSe in the Risky Industries. The main point is that, actually, the overall "context of the things" is totally different there and it cannot be used to explain the BSe in Risky Industries as NNT uses in his book.

First, in the Stock Exchange areas all of the BSe are caused by the Humans, but in the Risky Industries they are caused by Humans, Equipment, or Procedures (which provide interaction of Humans and Equipment).

Second, the point is that the employees in the Stock Exchange (investors and traders) are a particular type of persons (humans) who do their job with clear intention to earn some (if possible: A huge!) amount of money. Some of them are successful for short or long periods of time, but there, the knowledge (as NNT said) is not primary reason for success. The primary reason for success (or failure) is (intentionally or unintentionally) undertaking decisions based on data available which contain huge amount of Randomness and uncertainty. In other words, in the Stock Exchange, the humans intentionally undertake (un)justified risks with the intention to obtain a gain. They are assured that they can do it because the only consequences are: Losing the money or losing the job. And they think that they can live with these consequences.

In the Risky Industries, people cannot become rich as it can happen in the Stock Exchange. People there work for decent salaries and they are not keen to undertake any risks. Maybe, one of the reasons for that is the fact that the damage of the human lives, assets, and environment in the Risky Industries threaten the long-term recovery of the planet Earth.

Third, having in mind that BSe, by definition, never happen in the past, the solutions how to deal with them must be different than that for solving other problems. This is something which applies to the internal organization and discipline inside the companies. The company must diagnose the BSe as soon as possible and the understanding of the specifics of the event must be reasonable. The particular action which will be directed towards the causes or (if the cause is not determined) towards the consequences must be implemented. History has shown that dynamic companies could provide better results than static ones.

In this area, there is another paradox: We try to implement the SMS in handling safety problems, but in regards to the BSe, where there is no knowledge about it. A missing knowledge could not help.

It is known that companies which do not have strong organization and good SMS are more successful in handling BSe. In this companies, the employees very often must improvise and these improvisations "feed" their creativity. Only creative approach could work with the BSe immediately when they happen. Companies with strong management are not so creative because, there, everything is subject to rule and procedure. In addition, the companies with weak organization have employees who are better trained for "weak signals" in regards the adverse events. They have instinct that something can go wrong immediately as they notice irregularity and it can help to register the BSe early.

Fourth, I can mention that there is a Regulatory requirement in the Risky Industries to implement a Safety Management System. Such a Regulatory requirement does not exist in the Stock Exchange, but most of the companies there are implementing something close to it: Risk Management. There is even the ISO standard covering areas of finance, business and economy: ISO 31000 (Risk Management). Anyway, "the context of the things" is quite different, in accordance of that what I mentioned as second difference, previously in this same paragraph. The risk calculated in the Stock Exchange is about the losses and it is based on data and some vague understanding of human behavior from the aspects of trading. And it is not so successful. Even the US Government Enquiry Report on Lehman Brothers stated that one of the reasons for bankruptcy was an obsolete Risk Management System.

The Stock Exchange's Risk Management is mostly based on the experience of the employees in the company. There, everything is secret. In the Risky Industries, it is quite opposite: There, everything is public by the regulation!

In the Risky Industries, the Regulatory requirement is established to disseminate the information about any safety-related event (not only for incidents and accidents) and to disseminate any preventive and corrective measures, not locally, but globally. The reports of any incident and accidents in the world are freely available on the website of any international Regulatory organizations in any Risky Industry.

The Risk Management, which is used in Risky Industries, is quite different and considerably more successful than the one in the Stock Exchange. There, managing incidents and accidents is done in a systematic way, through regulation for implementing, monitoring, and maintaining an SMS.

Fifth, I can mention that, although the Safety Management is based on probability (as many of the safety people like to say…), there is no probability, just likelihood (frequency) as explained in Section 2.3.3 (Analyzing the Event Later, We Realize that It Was Logical to Happen…). So, "the context of the things" applies also in this area: All these problems with the Normal (Gaussian) and "Fat-tails" distributions in the Stock Exchange could not apply to the Risky Industries. There, they are just neglected by the simple choice on how to calculate the risk by using matrices.

At the beginning of this century, with the regulatory requirement for introduction, monitoring, and maintenance of SMS in aviation, the intentions were to change the previous safety attitude from "reactive" to "proactive".

And it worked!

If you look at Figures 4.1 and 4.2, it will be clear.

Statistical data for all aircraft accidents and all fatalities in these accidents for the period from 1942 to 2019 in the civil aviation field are presented in these figures. With gray rectangles on both figures is marked the time period from 2000 to 2019.

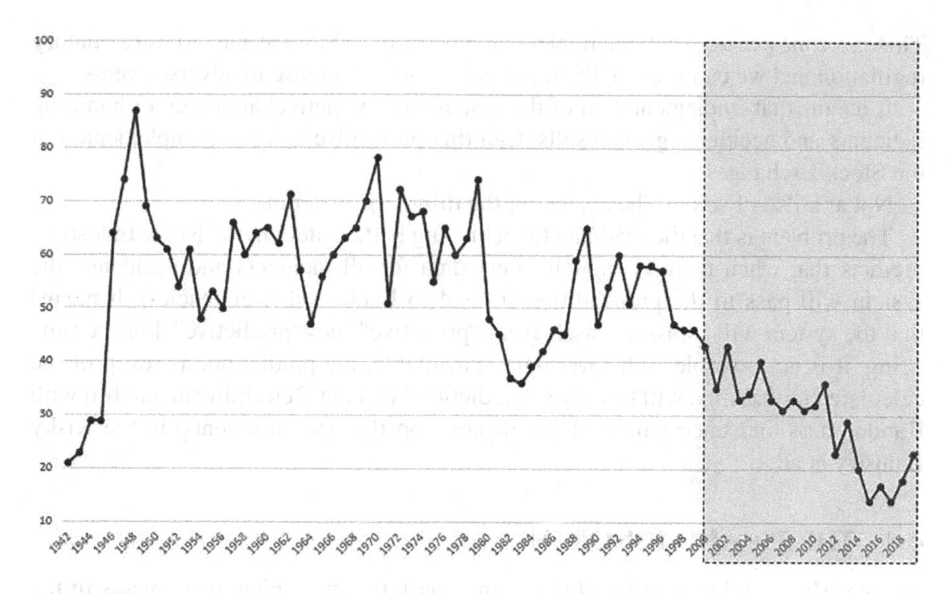

FIGURE 4.1 Statistics on all aircraft accidents from 1942 until 2019. (Data for producing these figures is used from http://aviation-safety.net/statistics/period/stats.php?cat=A1.)

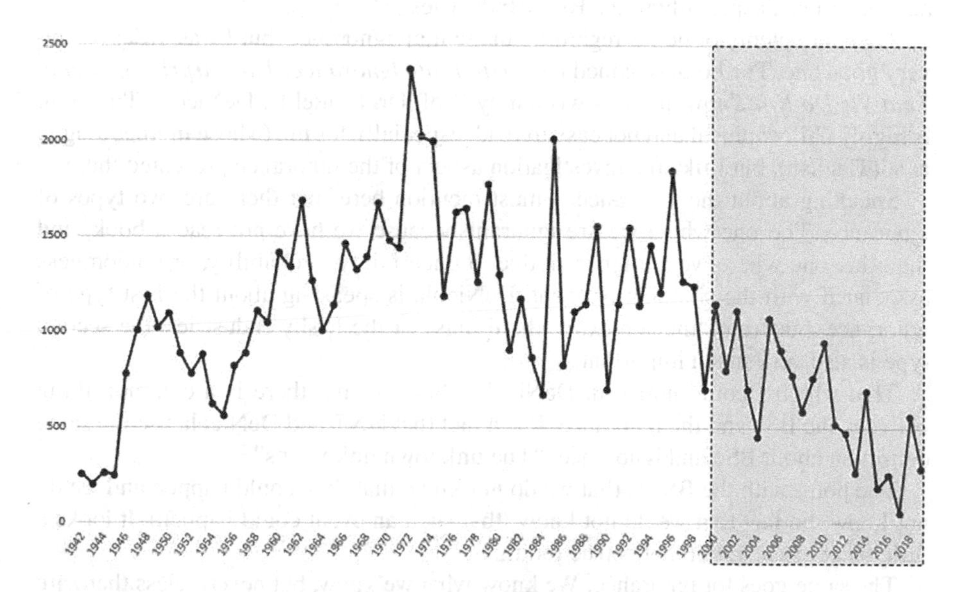

FIGURE 4.2 Statistics of all aircraft accidents from 1942 until 2019. (Data for producing these figures is used from http://aviation-safety.net/statistics/period/stats.php?cat=A1.)

This is a time period when the implementation of the SMS was passed as regulatory regulation and we can see that there is a tremendous decrease in adverse events!

It means that implementation of the systematic (proactive) approach of handling incidents and accidents gives results. Can this preventive action be implemented in the Stock Exchange?

Not at all! As I said: "The context of the things" is different!

The problem is that the roadmap for achieving better safety in the Risky Industries predicts that when there is enough safety data for all incidents and accidents, the system will pass to the probabilities, instead to likelihood (frequencies). It means that the system will transform itself from "proactive" into "predictive". For the time being, it is not possible. Whatever data is available, the prediction, as result of the calculated probability, will be only a "prediction" without Determinism and full with Randomness and uncertainty. This will not stop the BSe (accidents) in the Risky Industry at all.

4.4 THE BSe AND IGNORANCE

As one Islamic scholar had said (long time ago), the three biggest problems in the World are: The discrimination, the poverty, and the ignorance. The adverse events do not discriminate humans, assets, or industries; the poverty is not so spread in the industry; but the ignorance should be treated as most important factor in dealing with adverse events, especially in the Risky Industries.

There are plenty of books regarding the human ignorance, but I was lucky to find very good one. The book is named *Understanding Ignorance: The Surprising Impact That We Do Not Know* and it is written by Prof. Dr. Daniel R. DeNicola. The book is highly philosophical and not easy to read, especially for me (whose mother tongue is not English), but I like the investigation aspect of the ignorance presented there.

Speaking about the ignorance, I must mention here that there are two types of ignorance. The one where we are ignorant because we have not read a book, and the other one where we are ignorant due to uncertainty, probability, or randomness associated with the situation or event. DeNicola is speaking about the first type of ignorance, but from "the context of the things" of the Risky Industries, the second type is also very much important.

That which I could notice in DeNicola's book is that there is a common thing between the BSe and the ignorance. It is a fact that NNT and DeNicola use the same definition about BSe and ignorance: "The unknown unknowns".[2]

The point with the BSe is that we do not know that they could happen and we do not know the fact that we do not know (that such an event could happen). It looks a little bit confusing, but that is true for all.

The same goes for ignorance: We know what we know, but nevertheless there are things which we do know that we do not know, there are also things we do not know that we do not know. Even in the areas where we feel ourselves experts, there are still

[2] Actually, long time ago, the USA former Secretary of Defense Donald Rumsfeld used these terms first time in a press conference and these words became public. But, to be more precise, the first who made this classification was philosopher Ann Kerwin, approximately a decade before Rumsfeld's statement.

things which we do not know (that we do not know). This is something which happens very often with the humans: They think that they know something, but when they start, they notice that the things are different then assumed. This can happen very often during a test assessment in the schools. The students came ready for the test, but during the test, they realize that they are not so ready...

Anyway, having in mind that there are things which we do not know that we do not know, it means that we cannot measure them. Not measuring them means that we cannot quantitatively investigate them and this is a problem with the BSe, as well as with the ignorance.

Our knowledge could be described as a room with window. Having this in mind, we can describe our knowledge as perspective which we can see through this window. Whatever is seen through the window, we know it, but if there is mountain seen through this window, we do not know what is behind this mountain. To see what is behind the mountain, we need to leave the room with the window and climb the mountain. Or someone else should do it for us (the teachers) and then they can show us pictures of what is behind the mountain. Of course, we can use our imagination to assume what is behind the mountain, but this is also something which can be only a guess. And I have explained previously that we describe a guess by the use of probability.

This equality between the BSe and the ignorance can help us (maybe) to provide some measures to fight the BSe on the same way as we fight the ignorance: By learning! In the context of the BSe in the safety area, the important thing is, during this learning process, to be open-minded and to be ready to expect unexpectable.

Can you agree that everyone could see the pace of progressing of the technology? It is easy to be noticed because the new things caused by progress of technology are all around us. The point is that what you see is not always what is it. You can read more about in Chapter 7 (The Invisible Gorilla).

The point is that although the technology presents the most visible manifestations of the change of our ordinary lives, the problem is actually what is going on with our social understanding of other humans. The real question is: Is the technology helping us to become better humans?

The learning is good, but there is also the question what to learn to deal with the BSe in our case? Which type of knowledge would make a difference in process of stopping the BSe?

Of course, every knowledge will not provide solutions about the BSe, but it must be localized in the areas where the BSe can be expected. Knowing the activity, process, and operation in details, associated with particular capability to think in advance and particular creativity and open-mind, obviously can help in dealing with the BSe. Being conservative, but not paranoid, is also a capability that could bring benefits.

In the "equation" regarding the knowledge, the biggest unknowns are humans, and this is something which needs attention. Dealing with humans is the subject of each management system, and this is a simple explanation for why the Risky Industries needs regulation requirement to implement SMS.

One of the many problems with the humans is that, in general, they are reluctant to changes. Even if the change can bring big benefits to them, they prefer to stay in the same position or same environment. As DeNicola would say, the humans prefer "cognitive comfort" to "encountering the unknown". The humans are (intuitively) aware that the change can trigger new hazards and new risks for their wellbeing and they do not like it. Some of the humans think that what you do not know will not hurt you, but in the Risky Industries this will not work. This is the reason that in each SMS, the Change Management and additional Risk Assessment of the change are very important activities.

The Change Management is based on clear identification of why the change is triggered (regulation, economy, Human Factors, etc.), how the change will be executed (phased approach or abrupt approach), and what are the benefits that the change will bring to the process, company, or employees. As it can be noticed, humans do not like the change due to the ignorance (uncertainty) which is associated to the future situation (after the change is implemented), and the Change Management is, actually, fighting this ignorance.

Additional Risk Assessment of the change in advance is regulatory requirement and it must be the subject of the particular analysis of the newly introduced hazards and risks. There is nothing strange with the fact that the change will introduce new or novel hazards and risks or change the already established ones. There is no perfect solution to any problem, and there is always a cost which needs to be paid. This kind of "cost-benefit" safety analysis should tell how much the previous problems are solved and how many new hazards are introduced. With the new hazards, the emphasis should be put on their risks: Are they bigger and how much they can be eliminated or mitigated? The level of the analysis will depend on how big is the change and depending on that, sometimes it will need small analysis and sometimes the Safety Case[3] must be produced. Anyway, the change must be properly analyzed from the safety point of view and the appropriate actions must be undertaken.

There are few other aspects of the ignorance that cannot be neglected...

The ignorance is highly dependent on the environment. We try to build the picture of ourselves using others. Or better to say, we compare ourselves with others, our achievement with other achievements. In the sport, it is easy for me the reach a guy who is best in my school than reach a guy who is best in my city (country, continent, world...). Being best in the school in any sport is fulfilling until I do not go to city (country, continent, world...) championship. There, I can be "shocked" by the numbers in my sport achievements compared to others.

This is the same for the knowledge and the ignorance. The environment to reach the particular level of knowledge is dependent on the place where the knowledge should be achieved. Studying at Cambridge (or MIT, Oxford, ETH, etc.) is not the same as finishing in other universities in any country. There are environments which will provide conditions for better results, and they will push each individual to give more.

The same is with the Risky Industries. In the countries where there is considerable safety culture, the chances to provide safety are better. And this is the reason that building the particular safety culture is primary task of the Regulatory Bodies in each country!

The next aspect, which is in the favor of the ignorance, is that we (humans) feel safe in ignorance. Not knowing the bad things that could happen to us makes us

[3] In aviation, the Safety Case is a structured document where the arguments provided by Risk Assessment are presented to the Regulator that the change will be safe when implemented.

(unconsciously) feel good. But we need to be cautious with this statement. As I mentioned earlier, it is an unconscious feeling: If we know that something bad could happen to us, we will try to eliminate the risk and undertake actions to protect ourselves or to mitigate the consequences.

As another aspect, I can mention the situation (especially for managers) where the humans are ignorant about some process (situation, event, system, etc.) or they must make decisions about something, but there is no enough information to make a good decision. Obviously, the ignorance is present here and it will affect the decision-making process. It means, it is a problem for humans to make decisions regarding something, if not known or under high uncertainty. This is a part of the Human Factors and it creates anxiety for the managers. The point is that usually the decision under such a circumstance is always skewed in the direction to the areas where there is some (mostly) personal benefit. Very rarely is the decision skewed into direction of areas where consequences will be minimized.

Usually, the managers are those who afraid of making a wrong decision about how to fix something important, especially, because they could be blamed for doing the wrong thing. The fear of blame makes the managers not to be so brave to try to solve the problem. In many cases, when they have to make the decision about what to do, they will try to find a "save haven" by transferring responsibility to someone else to make the needed decision. Unfortunately, they do forget that the accountability still stay with them and it cannot be transferred.[4]

4.5 THE BSe AND PROCEDURES

As explained previously, there is one very important factor with the BSE and this factor is: Humans. Safety investigations showed that the biggest reason for the adverse events in any industry are humans and their imperfection. To improve the humans, there are many ways, and plenty of them are covered in this book. This is the section where I will speak about BSe and their connection with the operating procedures.

The procedures are simple written rules (instructions) how to deal with the processes. They are the backbone of every management system and there are two types of the procedures: System Procedures and Operating Procedures.[5] The System Procedures apply to the management system and the Operating Procedures apply to the production process (operation).

By following a particular Operating Procedure, trained, skilled, and educated employee (humans!) will produce a product with good quality. But this happens in ideal world. Sometimes, due to different reasons, the product could be scrap. One of the biggest reasons for this to happen is not following the steps in the procedures by the employees. Let's explain it in more detail…

The pilots do not wonder how to fly the aircraft because there are procedures on what to do during the flight. They are trained to follow the procedures which (depending on the type of the aircraft) are very similar to the procedures used globally in the aviation world.

---

[4] More details about responsibility and accountability of employees are provided in Chapter 10.

[5] These are known also as SOP (Standard Operating Procedures), but I prefer to use just Operating Procedures in this place. It just seems to me that the word "Standard" is unnecessary…

The statistics says that the event which is uncovered by procedure or the event which happened as a result of missing step in the executing of the procedure could happen extremely rarely in the aviation sector. In the cases where something happens and the pilots miss a step in the procedures, they will find themselves in unknown area. They will not know what to do because they were not trained for such a situation, so they have to improvise. The same statistics says that the probability that the pilots will do something terribly wrong during this improvisation will rise by approximately 30 times, compared to the normal operation. The next mistake/error in the improvisation (if the first improvisation was wrong) could happen with the probability of being 80 times bigger compared to the normal operation.

This is actually the BSe!

Not following the procedure or missing the step(s) from the procedure has all characteristics of the BSe. The BSe will happen and it will be a shock for employees. They will panic, and things can go from bad to worst in a matter of seconds. Do not forget that the panic is contagious: It can be extremely easy transferred to other people.

That is the reason that during training provided in the Risky Industries regarding the Operating Procedures, the employees must be trained also about the bad things which could happen if the step is missed or the procedure is not followed. This is one way where knowledge could make a difference: Knowing which bad thing could happen and what are the consequences if it happens, could prevent neglecting the procedure or missing a step from the procedure by the employees.

In my humble experience, the training regarding the procedures was missing in many companies where I have worked. In general, I have not met company providing training about the adverse events and their consequences that could happen if the procedure is not followed as it is written. If you think, this is not necessary, then allow me to give you a small but important example: The Virgin spaceship accident.

The Virgin Galactic is an American/British company developing a commercial spacecraft (the newest version is named SpaceShipTwo), but not for the research of the space, for business purposes. They intend to use this spacecraft to provide suborbital spaceflights to tourists. Maybe later, they will be used for suborbital launches for space science missions. There is a specially developed airplane which is used as carrier of SpaceShipTwo, known as White Knight Two. The White Knight Two, carrying the SpaceShipTwo, takes off as an aircraft and at an altitude approximately of 14.5 km the rocket motor of SpaceShipTwo is ignited. This ignition of the rocket motor in the SpaceShipTwo will bring the ship to altitude of approximately 22 km.

In October 2014, the accident happened during the test flight when the space ship disintegrated. One of the pilots died and other was seriously injured. The investigation showed that the tail of the space ship was deployed too early and the stress caused on the tail actually contributed to the disintegration. The flight procedure strongly requested to have the speed of, at least, 1.4 Mach for deployment of the tail and pilots did it at 1 Mach. If the pilots were told that, the early deployment of the tail will cause the disintegration of the space ship, they would not do it. Obviously, they have been told that they may not do it later, but not early...

5 The Gray Rhino Events

5.1 INTRODUCTION

On my regular visits to the "Borders" bookshop in Abu Dhabi, I had a chance to find a book named *The Gray Rhino* from Michele Wucker. The subtitle of the book was "How to Recognize and Act on the Obvious Dangers We Ignore". The resemblance of metaphor "Gray Rhino" with metaphor "Black Swan" was very obvious, so I spent some time on the book. That time was enough to make me to buy the book and to read it thoroughly.

It is a book issued by St. Martin Press in 2016 and it primarily deals with the events in macro-economy, sociology, and politics with a move (later in the book) to the area of micro-economy and everyday human lives.

As it is written in the book, the Gray Rhinos' events (GRe) are hazards with particular risks (highly probable to happen), which can be easy predicted and which can produce very high impact. The problem with GRe is that, nevertheless, they are associated with some signs (clearly noticeable) that tell us they will inevitably happen, but the humans do not undertake measures to eliminate or to mitigate them. As Ms. Wucker says, the GSe are children to the "elephant in the room" (clearly noticeable) and the improbable and unforeseeable "black swans" (huge, usually bad impact).

The GRe are not random events and they happen very often all around us, all the time. Having in mind that they are highly probable, it makes them very much predictable, but this is not the problem…

The main characteristics of the GRe is that they happen after one or a series of warning events (which can be clearly noticed), but these warning events (harbingers!) are totally neglected by the humans. Ms. Wucker's book investigates the possibilities that the GRe are consciously neglected due to human characteristics (maybe it is natural?) or simply by denying that they exist. If we deny them, we automatically neglect them. In the cases when we do not neglect them, we usually process these harbingers in a totally wrong way, so the outcomes of the processing made us to neglect them. This is actually a part of the wrong decision-making process and, mostly, the decision-makers are prone to cause these bad events. These are, actually, the guys who neglect the harbingers of the GRe.

There is an interesting connection between the BSe and the GRe. Even NNT in his book is mentioning few "harbingers" which were neglected and, with the time, they transformed themselves into the BSe. The difference is actually based on these harbingers: If the harbingers are not known as existing, at the time of noticing them or their connection with the BSe is not known, then we may speak about developing BSe. So, compared by BSE, the GRe are events where the harbingers are known, but anyway, they are (intentionally or unintentionally) neglected.

Roughly, the book can be divided into two parts. The first part elaborates the GRe from different aspects and I really enjoyed the reading. The second part is mostly

DOI: 10.1201/9781003230298-5

about what to do with the GRe and there are a lot of examples given from economy, politics, and environment fields. I did not enjoy second part simply having in mind that my interest in the GRe was not in these areas. Anyway, I can sincerely recommend to read this book.

5.2 THE GRAY RHINOS

Compared with the BSe, the GRe are events that have a particular level of uncertainty of happening, but this level of uncertainty disappears when you see the harbingers which warn you that the GRe is on its way to happen. Having this in mind, we can say that "the context of the things" with the GRe in the Risky Industries is that they are events which are known to the Safety Managers. These events can be found in the List of Hazards in each company, their risks are determined, but still, the company's employees fail to recognize or simply, they neglect them. Of course, it will create terrible consequences, when it does happen later.

The metaphor behind the name "Gray Rhino" is undertaken from the rhinos in Africa. They are massive and quiet animals that have poor eyesight and very good sense of smell. They can be a life-threatening hazard to you if you provoke them. Meaning of "provoking rhinos" is not to do something particularly which could be understood by them as danger. The rhinos can attack even if they notice you are coming close to them. Simply: They can feel endangered only by your presence in their vicinity. Plenty of people will go to safari and they will see the rhinos, but the wise advice to them will be: Do not to come close to them!

Specific danger from rhinos in the nature is: If only one of them attack you, you are faced with a mass of thousand kilos which is running very fast towards you and there is no "safe haven" from that. The car or SUV will not protect you and it is too late to realize that you need to be inside an armored military tank in that particular moment. So, there is no chance to stop the event and the consequences will be terrible. The rhinos in Africa are really dangerous animals and this example should warn you, that although the rhinos look quiet, the consequences from their attack could be catastrophic. Even the BBC mention them in their documentary series "Deadly 60", presented by Steve Backshall.

In the scope of safety investigation's aspects of the GRe in our everyday lives, these are events that happen almost every day. We notice the harbingers that something bad can happen, but we fail (intentionally or unintentionally) to handle them in proper way before they happen. Of course, this will result in a total disaster and only dealing with the consequences later, could mitigate the situation. This is actually the biggest difference between the BSe and GRe: The GRe depend mostly on the human decisions, while the BSe are totally unnoticed until they do not happen.

In the previous chapters I have mentioned that 9/11 events are great example for the BSe and there are plenty of examples for the GRe in the book of Ms. Wucker. Anyway, it is good to give here a more concrete example for GRe. As good (also very sad and very tragic) example, I would mention Steve Jobs...

Steve Jobs was diagnosed with pancreatic cancer in November 2003. The prognosis for pancreatic cancer is usually very poor, but Jobs told his colleagues that he has a rare and less aggressive type of cancer. There are different stories about that on Internet, but I read the book of Walter Isaacson and it looks to me very trustworthy.

Anyway, Jobs neglected the doctor's advice for a full nine months and did nothing. Actually, he was trying some alternative medicine (diets) and no one knows what he was doing. There are different comments from the doctors about this alternative treatment done by Jobs himself, but most of the doctors stated that his cancer was actually curable (with good chances to be cured), if he followed their advice for surgery. Some of them, even stated that his decision to try alternative medicine was an act of suicide.

Finally, in 2004, when his health had deteriorated a lot, he underwent a surgery and it was stated that the tumor was successfully removed. He did not receive chemotherapy or radiation therapy and everyone was sure that things will be alright.

But it was not alright…

In 2011, he was subject to a liver transplantation surgery, because the cancer had spread to his liver also. After the transplantation, his health situation deteriorated and he died in his home later that year.

This is a very sad story and I am not happy to use it here, but I simply cannot find a better example for the consequences of neglecting the harbingers of the GRe. Actually, this is a story that should warn us not to neglect (ignore) the harbingers simply because they happen on the small (individual), but also, in large (worldwide) scale.

In general, the message hidden in the GRe is: If you have some problem, do not postpone its solving. The problems are like children: They grow up and one day, they will be bigger than you and they will slap you very hard.

There are plenty of such examples all around us, and these examples can be treated as GRe:

- The GRe is an exam which students know the date when it will happen, but they, due to many reasons, just fail to start to preparing on time;
- The GRe is knowing that you have high cholesterol, but anyway, you are reluctant to start with regular physical exercises;
- The GRe is knowing that you have diabetes, but anyway, you are reluctant to start a recommended diet or use the appropriate drugs;
- Etc.

One of the best explanations of the GRe came from Mr. Warren Buffet through his "Noah Rule". Mr. Buffet is well known for his wisdom in the business and, from time to time, he will disclose to the public some of the rules which he applies in "doing business". In the annual report for 2001 of one of his companies (Berkshire Hathaway), he mentioned that he did not abide to the Noah Rule: "Predicting the rain does not count. Building arks does!". Actually, in this report, he confessed that some of the risks for that year were foreseen by him, but he did not do anything to eliminate or to mitigate them. The year 2001 was the worst year of the Berkshire Hathaway company…

Ms. Wucker provides categorization of the GRe in her book where, in a table, you can find description of eight types. I will not go in detail about any of them, and I recommend you to read Ms. Wucker's book for more details.

The point with all these GRe is that, although there are different reasons why bad things happen, the pattern regarding the GRe is always the same: Humans fail to recognize any of the harbingers for them! Ignoring these warning signals means that the GRe later happens and we need to deal with the consequences.

5.3 THE PHASES OF GRe

Roughly, there are few phases[1] of the GRe:

1. **Indication Phase** – This is a phase when one or few events happen as sign that the things could go differently than expected (the GRe could happen). The simple example is driving a car on the road. You drive a car and there are also a lot of cars on the road. It is a dynamic system. Anyone of these cars has its own potential for a crash with your car, but you (as a driver) monitor them and you adjust your driving on the current situation, maintaining your and others' safety. This is something that shall always happen during driving a car, but unfortunately in reality, it fails many times. One failure is that you notice that one car, far away from you, is coming in your direction. There are two possibilities in this phase: You do not notice the sign (movement of the car in your direction) or you do notice that and you neglect it (underestimating the danger of the situation). If you do not notice the sign, then it is really big problem. But if you notice the wrong movement, than you may enter the next GRe phase.

   The first phase is not so simple. In this phase, regarding noticing the harbingers, there are few very big problems. Imagine, there are plenty of people in the room, but only one would notice the fire ("gray rhino") in the building. He will rise an alarm, but how many of the present guys will believe him that he has seen smoke or fire? How many of them will wonder why they have not seen the fire? How many of them will question the reasoning of the guy regarding the seen fire? How many of them will think that this is a scam?

   So, by seeing a harbinger, it does not mean that an action will be undertaken...
2. **Neglecting Phase** – This is the phase when the warning signs happen in front of you, but you do one of these things:
   a. Simply you do not notice them! This is something explained by one other "safety animal" called Invisible Gorilla (IG) and explained in Chapter 7 of this book;
   b. You process them in the wrong way by using a wrong method, wrong procedure, or wrong tools/instrumentation for processing. This can

[1] These are phases which I have introduced for the purpose of explanation of happening of GRe in Risky Industries in this book. Ms. Wucker, in his book, is using different phases explained as stages: Denial, Muddling, Diagnosing, Panic and Action. There is nothing wrong with Ms. Wucker stages, but I think, the phases must be adapted to the nature of business in the Risky Industries and to the way how the safety is maintained there.

happen very much in the industry. There are plenty of examples of big engineering mistakes which have finished with accidents;

c. You have wrong assumptions what can happen. Simple example: During your driving a car you notice that car is coming in your direction, but you assume that the driver will rectify the wrong direction or he will press the brakes before he collides with your car. You may even press on your horn, but he does not do change car's direction (due to many reasons). The crash happens…;
d. You neglect them. Simple examples: Students and exams, athletes and coming matches (or championships); or
e. You just deny them. Simple example: Steve Jobs.
f. Etc.

3. **Happening Phase** – This is a phase when the GRe happens and there is nothing to be done to stop it. In the case of car coming in your direction, it is situation when the other driver does nothing and you are urged to press the brakes of your car. Unfortunately, it could be simply too late. In the scope of this example, one possible scenario is: The other car will crash into your car; it is a terrible crash; it is a devastating crash. There are casualties (I hope you are not dead!), the assets (cars) are ruined beyond repair, and that is it…
4. **Consequence Phase** – The GRe happened and now, it is time, as soon as possible, to eliminate (if possible) or mitigate the consequences. In my example regarding the car crash, the activities which follow are: The traffic on the car crash site must be secured and contained (to prevent crashes of other cars between them or, in already crashed cars); the emergency services and police must be informed as soon as possible; and the injured must be given First Aid, they must be transported to the hospitals, as soon possible. The police will start on-site investigation and other types of investigation (if needed) will be initiated. This is something which will happen immediately after the GRe happen, but in the long run, dealing with the consequences will need more time and more resources (money). In addition, recreating the previous situation could be very costly. And this is the paradox with the GRe: We neglect or deny the harbingers in the time when needed resources to eliminate the GRe are very low. But later, when it happens, the costs will be many times higher than dealing in advance. You must have in mind that "the medical care in emergency room is the most expensive care which the patient can receive". So. we should invest in preparedness and resilience.
5. **Analyzing Phase** – This is not necessarily a phase which belongs exclusively to the GRe, but it applies whenever something bad happens in the Risky Industries. This is a phase where the analysis must be done (usually by the Regulatory Body) with the intention to determine what was the root cause for the bad event? The finished investigations must address what caused the event, all things and reactions which were wrong, and the directives and recommendations that must be made (as legal means) so this does not to happen again in the future.

5.4 DIFFERENCES BETWEEN BSe AND GRe

There are few considerable differences between the BSe and GRe...

The first one and the most important one is that in the book of NNT, the BSe are explained as "unknown unknowns" and simple translation of this explanation is: We do not know that we do not know them. In other words: We do not know that such events are possible until they happen.

NNT, in his book, defines also other events and in the scope of above-mentioned definition for the BSe, he defines them as "known unknowns" and "known knowns". Let's remind you that these "knowns" become popular phrases by the speech of the USA former Secretary of Defense, Donald Rumsfeld.

By these definitions, the GRe are "known unknowns". Simple explanation of the GRe in the scope of this definition could be: We can notice them and we know them, but we do not know (consciously or unconsciously) to what they will develop until they do not develop. And it is too late then...

The "known knowns" are, in simple words, the basis of the today's Safety Management Systems when we identify ("know") the hazards (as "knowns"), we calculate the risks for each of them, and we provide (known) measures for elimination and/or mitigation to stop/mitigate the events.

In general, the BSe are unknown in the theory and also unknown in the practice, until they do not happen. The GRe are known in the theory, but their happening is neglected in practice, until it shows up and we cannot stop it anymore.

You may think that it will not happen in the practice in the companies of the Risky Industries, but I had a situation in a meeting where the cyber security was discussed and Head of Safety in the company said that there are plenty of defenses in the company, so he does not think that the cyber security is a hazard (?). Later, I explained to him (privately) that, already, there are data about plenty of cyber-attacks in the aviation companies with already established defenses, so these things may not be neglected.

The second difference between the BSe and the GRe is that the BSe are direct events that happen without any harbingers while the GRe are associated with harbingers. The GRe happen only because we neglect these harbingers by assuming that nothing will happen or because we do not recognize the hazards and risks hidden in these harbingers. Anyway, in terms of the harbingers, it must be recognized that if the harbingers are not recognized, then the GRe could become the BSe, depending on the situation.

There is also resemblance between the BSe and the GRe: Both of them, if they happen, nothing can be done to stop them.

But if we notice the harbingers of the GRe in advance, the measures to stop the GRe can be used as opportunity for considerable change of our system and, consequently, the changes could make it better. Such a simple example was Year 2000 (Y2K) problem in IT industry at the end of the last century. Let me "refresh" the older IT engineers and "educate" the new IT engineers who are not familiar with Y2K problem...

If we can agree that the computer era started approximately in 1960s, in that time the hardware was not so developed and the software was very much connected with this rudimentary hardware. In that time, the good programmer was a guy who could write software and his software will save a memory space. That was the reason that these years, the year in the dates in the software, was written only with two digits (instead year 1967, it was written as 67).

This worked for approximately for 30 years and in the 1997, the Sales Department of Boeing noticed that they cannot use two digits for year 2000, because their computer system would interpret it as year 1900. They were dealing with the new orders for their aircraft, which usually happen approximately 2–3 years in advance. They reported the problem and, in the IT industry, an alarm was raised: This can happen to any software program which is written before year 1997.

This was not a simple problem.

The real problem was that no one could predict how the software will react to this "bug" and what will happen after the New Year's Eve in 2000. Obviously, in the future, the computers will have a problem to cope with the year's expressions in the dates. This was extremely critical for the banking sector where all interest for the credits and deposits are calculated on a date basis.

The international community reacted promptly and this bug was used as opportunity radically to change the ways how IT hardware and software was produced. In addition, a lot of contingency plans were produced and back-up systems were implemented. This was a situation where the World showed that, if united, the problems can be eliminated or at least, they can be highly mitigated.

And all these activities worked!

There were just a few problems after the New Year's Eve of year 2000, because everybody was vigilant and prepared to solve possible problems. In general, all these activities were about "thinking for the future", so today's hardware and software is qualitatively based on the principles which are considerably different then the hardware and software principles from the last century. This is a beautiful example that, the GRe, if noticed early, can be opportunity to do things significantly better!

And here I can mention something which is very interesting…

The BSe and the GRe can be explained as one bad event that has two different outcomes depending from the sides in this bad event. In her book Ms. Wucker give an example of the Greek-Persian war which happened from 490 B.C. to 480 B.C. In this war, the Greeks won, but the point is that, for them, this war was a GRe. They had noticed harbingers of the Persian's intentions in advance and nevertheless, with less resources, their strategy was reasonable and successful. From another side, the Persians (their king Xerxes) were not so wise and their strategy was based on the advance in the number of humans and other resources which they abundantly had, so eventually, for them, their defeat was a BSe (totally unexpected and very much catastrophic).

Nevertheless, this is an example from history, it could happen even these days in a sport (team or individual matches and championships) and in the industry when some company goes (unexpectedly) into bankruptcy.

5.5 HOW TO DEAL WITH THE GRe IN GENERAL?

It is a normal question in the title of this paragraph, but the answers can differ from the approach of the Safety Manager in the Risky Industries. My approach, which I would like very much to recommend, will be: Sooner is better!

Having in mind that in Section 5.3 (The Phases of GRe) where I have explained the phases of the GRe, let's see how we can eliminate or mitigate the GRe and their consequences in particular phase:

1. **Indication Phase** – "Sooner is better" means not to lose a chance to deal with the GRe in this phase. This is the phase where the GRe can be eliminated (stopped) in total! It is essential to register and analyze the harbingers. If the GRe cannot be eliminated, this is a good time to prepare to eliminate and/or to mitigate their consequences. As it can be seen, there is a need for a five-subphase process:
   a. The first subphase is to implement good monitoring system (24/7) in your company with competent and experienced employees. These employees must be capable of interpreting all harbingers in the right way and they must do everything to gather more data about these harbingers;
   b. Not necessarily will these employees need to analyze the gathered data, but obviously, for the second subphase, there is need for other employees who will have knowledge and experience to do that. A good analysis will provide hints what type of measures (corrective and/or preventive) needs to be established;
   c. The third subphase can start by producing corrective and/or preventive measures on how to deal with the incoming GRe;
   d. The fourth subphase will be execution of this measures: and
   e. The fifth subphase would be hindsight analysis regarding "lessons learned" and possible changes into the established QMS or SMS, which could provide better handlings of the GRe in the future.

   Of course, during all these phases, the monitoring of harbingers must be maintained, because I am speaking about dynamic environment and any change must be reported or maintained in a timely manner.
2. **Neglecting Phase** – After registering the harbingers, there is a possibility that they are neglected due to wrong results of the analysis. To deal with this phase, it is not enough to maintain continuous monitoring with intention that harbingers will show again. They will maybe be present all the time, but they will be neglected as signs of the GRe. Having in mind that we are living and working in dynamic environments, may be the harbingers will pop up again in changed (evaluated) form. It means that they will not be recognized as already analyzed harbingers, but they will be recognized as novel ones. If the monitoring is maintained, it can help to deal with the harbingers. There is not too much to be done in this phase…
3. **Happening Phase** – This is a phase when the GRe is ongoing and only the resilience of the company and the speed of reaction (under assumptions that there are measures already establish what to do) could help to prevent a damage. Here, the human resilience is more important simply because most of the humans will panic during unfolding of the GRe. In such a situation, it will not help having in advance some back-up systems and contingency plans with already established procedures of what to do. Improper response from employees due to panic could spoil the reaction on the GRe happening.

It is very much important to understand that, the back-up systems, execution of emergency procedures, and contingency plans, in this phase, could make a difference! This is place where well-established SMS and "prepared in advance" employees should show their effectiveness and efficiency.

4. **Consequence Phase** – This is the phase where the GRe already happened and, now, the focus is put on the elimination and mitigation of the consequences. Everything done in this phase must be fast and based on a reason. This is a phase where back-up systems and contingency plans with already established procedures should be activated as soon as possible. Anyway, the monitoring of overall operation and processes in the company must not stop in this phase, because new "animals" (bad events) could pop-up. Focusing on consequences and not maintaining ongoing processes (if there are still ongoing operations), could be fatal. The maintenance of other operations and processes in the company must be included in the back-up and contingency plans.
5. **Analyzing Phase** – This is a phase when (ideally) independent and impartial team of experts will investigate what just happened and how it happened. All aspects of the GRe must be taken into consideration and (very important) all results of the investigation, assembled as official report, should be disseminated to the companies in the same Risky Industry in the World. Usually, it is done through the Report issued by a particular international regulation body for each Risky Industry.

6 Specifics of the GRe in the Risky Industries

6.1 INTRODUCTION

The main question of Ms. Wucker's book is: Why we are not capable of recognizing the obvious hazards before they materialize?

She tried to respond to this question regarding the areas of economy, politics, and business in her book, but in general, there are complex reasons that can be connected as answers to this question. Some of them are part of the strategy and tactics inside the company, and some of them are part of the human nature. This second part (about human nature) is pretty much emphasized in the areas of the Risky Industries, simply because there is a general strategy followed by tactics, to earn a profit and not to endanger humans, assets, and environment.

In Chapter 5, the GRe were mostly explained as per the book by Ms. Wucker, and it was dedicated to the general GRe in the areas of economy and business. Even as such, it is clear that they cannot be neglected as factors for the adverse events in the Risky Industries The reason is simple: If the BSe are connected with the ignorance of the nature of possible adverse events or the GRe are connected by misinterpretation of the harbingers that they may happen, it can happen also in the Risky Industries. Of course, do not forget: In the Risky Industries, the consequences of GRe will be terrible!

As I have already explained in Section 2.4 (The Characteristics of the BSe in Stock Exchange Area), in the Risky Industries there are incidents and accidents. The incidents must never be neglected, because if you do not handle them, "the way for happening of accidents is paved".

The fact is that the GRe can finish with disaster and, as such, they must be subject of investigation in the Risky Industries. The worst thing about the GRe is to do nothing when the harbingers are noticed. This is very much important to the economy, politics, and business, but for the Risky Industries, this is amplified many times.

The main point with the GRe in the Risky Industries is the fact that they are (mostly) connected with human behavior and, as such, they are part of Human Factors (HF). HF is a scientific discipline dealing with analysis of reasons for non-intentional human behavior resulting in errors and mistakes.

Of course, there are also few other things and they will be the subjects of discussion in the following paragraphs…

DOI: 10.1201/9781003230298-6

6.2 HOW TO PREPARE TO HANDLE GRe IN THE RISKY INDUSTRIES?

Some would say: The GRe shall not happen in Risky Industry! The employees are trained to deal with all these things and they will notice the harbingers, understand what is going on, and eliminate or mitigate the causes for the adverse event.

I agree with all these things, but the GRe happen in the life, so are the Risky Industries part of the life…? Whatever is the answer, the GRe happen everywhere. That is the reason that we need to prepare.

To prepare for the GRe in Risky Industries, there are two questions which must be answered:

1. Is (are) the harbinger(s) of GRe noticed? and
2. Is (are) there any action(s) undertaken?

The answer of first question could be: "Yes" or "No". If the answer is "No", then we can rely only to our already established (general) measures for resilience of our SMS to protect our company, lives, assets, and/or environment. Obviously, the incoming GRe was not recognized and there is nothing else to be done in such a case…

The point here is that, later, with hindsight, we need to analyze why these harbingers were not noticed. If they were not known, then they are BSe. But if there is some other reason, clearly, it will show a lack of effectiveness and efficiency of the established and implemented SMS.

If the answer to the first question from above is "Yes", then we move to the next question which could provide (again) two answers: "Yes" or "No". "Yes" means: The noticed harbingers produce concern which will trigger an analysis in the company, and the outcome of the analysis will result with particular preventive or corrective actions which will be immediately established. "No" means that no action will be triggered despite the harbingers being noticed. And this is (again) a concern for an analysis…

There could be plenty of reasons why no actions were prepared and executed, but the first thing which needs to be clarified is: Why the harbingers of GRe were not analyzed if they were noticed?

And this is subject to a more thorough analysis. Later, I will mention few reasons for that…

Going further, to find a remedy for the GRe in the Risky Industries, there are two things which must be starting point for any measures or activities in the companies. The Safety Managers must take care in advance to:

a. Understand the GRe in general; and
b. Understand the GRe in "the contest of the things" in their organization (company).

Both of these things are connected with the Humans, the Safety Culture and the human performance in the company.

6.2.1 Understanding the GRe in General

Regarding the understanding GRe in general, I can give a simple example from our everyday lives…

When we feel ill or unwell, we go to a doctor. We go there because this is the person who is educated about our body and he is educated and knows how to determine our problem and what can be the remedies for the problem. His considerable level of education should produce knowledge which, associated with his personal skills and experience, should produce good results regarding our health problem.

The same thing applies to the GRe in the Risky Industries: The Safety Managers must get familiar with all phases of GRe and they need to be analyzed in advance. "Prevention" is always better than "Correction" and that is the reason why we produce our SMS: The first thing is to prevent by eliminating and mitigating the risks and, if it is not successful, then we strive to eliminate or mitigate the consequences.

Regarding the GRe, it is wise to be more conservative in assumption and in processing the data. For someone, it is maybe not economical, but I do believe that this "conservatism" will provide more benefit than damage.

6.2.2 Understanding the GRe in "the Context of Things" of Your Organization

Each company differs from others. Even in the same industry, the company differs in:

a. Volume (of premises, resources, production, etc.);
b. Humans employed (education, culture, religion, social status, etc.);
c. Equipment (used for production/services or other processes);
d. Procedures (operational, logistic, financial, system, etc.);
e. Environment (continent, country, city, etc.);
f. Etc.

All these things produce differences in the established management systems inside the companies. It means that "the context of the things" is different everywhere. And this is very much important to understand: The same incidents and accidents have different "stamp" depending on the companies where they happened.

So, it is not just enough for the Safety Managers to get familiar with the GRe in general. They must get familiar with "the context of the things": what to do if the GRe happen in their company. This should not be used only for prevention but also for elimination and/or mitigation of the consequences, if the GRe happen in their company.

6.3 Information vs Data

An important thing regarding the GRe is how they start. At the beginning, there is some piece (one or more) of data or information which is actually harbinger of the GRe. If we **notice** it and if we **processed** it in the right way, it is a warning sign which will trigger action to prepare or execute the preventive or corrective measures to handle the GRe before it happens.

But let's speak a little bit about data and information. Are they the same thing…?

Although, in the ordinary life, we mostly took data and information as synonyms, they are not the same thing…

The data are usually the facts which are used to obtain or to produce an information. The data can be something which we notice, see, feel, measure, etc. The data alone rarely produce some information. The data need to be organized, structured and, as such, processed to provide valuable information. And how the processing of the data can be done is roughly explained in Section 2.3.3 (Analyzing the Event Later, We Realize that It Was Logical to Happen).

So, there is a science called Statistics (part of Mathematics) for processing the data and there is science called Theory of Information, which is dealing with transport and processing of information. Usually, the Theory of Information is built up on the outputs of the Statistics.

The main point is that not always can the processing of data (if available) produce valuable information. The random data can produce only probability, and I have explained what is probability in the previous chapters. But although the availability of good data can produce particular information, keep in mind that it is the humans who process the data.

There is one beautiful example about processing data from First World War. At that time, many of the aircraft, after the air battles, survived and they came back to their domestic airports. The holes of the bullets from hostile aircraft could be found on their fuselage and their wings. The technicians, noticing these holes, gathered the data about the holes' positions and those data were sent to the manufacturer of the aircraft. In those manufacturing companies, the engineering staff concluded that these are places which needs to be reinforced with stronger material and they started to do it.

But there was one clever engineer who pointed to something that was quite different than what his colleagues were considering. He said that these data, about the holes from bullets are processed the wrong way. The places which needed to be reinforced are not those where the holes from bullets can be found, simply because if the aircraft came back, these holes were not critical holes. The places which needed to be reinforced are those where you cannot find a hole from a bullet because all the aircraft which did not come back were obviously hit in these places. It makes sense, correct?

Another good example about the wrong conclusions after processing the data is one where the fake news is used. There was a documentary regarding "fake news" on one of the TV channels which I had chance to watch when I was writing this book. An information was presented there that it was noticed that in Latvia (Europe) Nazism is increasing. The information was gathered from the data of downloading the Hitler's book "Mein Kampf" from one of the Latvia's book websites. But going further into data, it was noticed that 70% of the downloaders were unregistered users. The data for downloading the Harry Potter books were, on the contrary, from 70% of registered users. Comparing the two information, it is easy to understand that downloading Hitler's book from these users was scam to provide information that Nazism will prevail in Latvia in the future. And this is actually an act of "Special War" or something which is called "Cyber War". So, the number of downloading of the "Mein Kampf" is not a harbinger of increasing the Nazism in Latvia (it is not harbinger for GRe at all).

There are many interesting and simple examples which I can point to the case of correlating data from different areas which sometimes can give very funny situations. Writing my second book I went on Internet and I wrote in the search engine "funny statistics". Approximately one million sites were found and checking few of them was a really fun. In one of them was expressed the graphs of statistical data regarding selling margarine in USA and the rate of divorces per month in Maine (USA). These two events are totally independent of each other and, by the theory of probability, they cannot affect each other. I must agree that the graphs were almost identical, but let's be clear: What the hell does the consumption of margarine in USA have to do with the divorce rate in Maine (USA)???

Do not forget that the value of information is always associated with particular uncertainty, so again, we use probability to value integrity of data and integrity of decision.

The reasons for all these things are that especially in the cases of the GRe, there is not so much data which can be recognized as harbingers, but some of these data can clearly point to the coming GRe. Anyway, it is all about the humans, their processing abilities, and their decision-making capabilities.

Speaking about human's "decision-making capabilities", we must emphasize that there, the compromise is very much present. The compromise, in the case of the GRe, can be defined as underestimating or overestimating some of the risks when we undertake decision-making. The problem with the compromise is that, it could be very much biased, and this happens very often in the Stock Exchange, the economy, and the politics. The people there very often undertake particular risks for the sake of the particular gain, which means that they underestimate the risks. In the countries of Balkan Peninsula in Europe, there is adage: You do not play with fire – you will not be burned! This is a pretty good example of overestimated risk. It is not the point with this adage not to use the "fire"! The point is to use it prudently and in a clever way!

So, the GRe are strongly connected with the harbingers which can show up as bunch of data (which we process to get the information). If you go back to the third sentence of this paragraph, you can notice that there are two bolded words: "notice" and "processed".

It means that there are two steps how to deal with the harbingers. The first one is to **notice** them, and the second step is to **process** them. Both of them have the same importance, but this can happen only if they are executed one after another. Obviously if we do not notice harbingers there is nothing to be processed, which gives more value to noticing the harbingers than to processing information about them.

A noticing of something is an individual characteristic of the humans, and it can be improved by training or spoiled if you (generally) do not care. This is something which I will cover speaking for the Human Factors (HF). From the point of the GRe, I would pay more attention to processing the data (harbingers).

The processing of the data to provide an information is connected not only with the methods of doing it, but also with the skills to recognize the real information hidden into the data. Of course, bigger experience of the "processor" will provide better skills.

6.4 DEFENSE LINES FOR GRe

Roughly, there are two lines of defense for the GRe: The company and the Regulator. Both of them will be explained in detail in the next two paragraphs.

6.4.1 Defense Lines of the Company

The companies in each industry (Risky Industries included) do business. By "do business", I mean: The humans establish a company to earn to themselves a profit which they will use later to maintain and improve their lives. So, the primary point in "do business" is to provide a profit. Each accountant knows that if the costs of the business increase, the profit will decrease. And this is something which is very much used by most of the companies: Increase the profit and decrease the costs!

You will say: There is nothing wrong with that, and I can agree!

But there is something wrong in the methods used to decrease the costs…

Let me speak about one simple example…

In my humble beginning in the Quality and Safety areas (around 2005), my learning process showed that companies in the Risky Industries are very good in calculating costs of measures used for preventing incidents or accidents, but they fail in calculating the costs of benefits which this prevention could bring to the company.

Established in 2003, there was a White House (USA) Commission for Flight Safety, as a response to the complaints of airlines regarding the Regulatory requirement to establish Safety Management System (SMS). Actually, not only the airlines complained but this regulation was also an economic burden for other subjects in the aviation (ANSPs, manufacturers, MROs, airports, etc.). As a minimum (dependent on the size of company), any company needed to employ at least one more person: A Safety Manager.

The White House decided to investigate the possible benefits from implementing SMS in the aviation and there were few findings in the report:

> *The decreasing of 73% of safety risks will bring to the airlines 620 million US dollars' savings every year. Every safety incident (compared by the number of flights) cost aviation subjects 76 US dollars per flight. By implementation of only 46 recommended safety improvements these costs decreased to 56 US dollars per flight.*

Anyway, this was not understood by Boeing in the last few years…

The race for profit expressed by the wish to shorten the time for testing of the novel Boeing 737 MAX aircraft cost the lives of 364 people.[1] The damage to Boeing's

---

[1] On October 2018, the Boeing 737 MAX aircraft owned by Lion Air crashed in Indonesia and 6 months later (March, 2019) another same type aircraft (owned by Ethiopian Airlines) crashed in Ethiopia. The cause of these accidents was malfunction of the embedded Maneuvering Characteristics Augmentation System (MCAS) in the aircraft which needs to correct the stability of the aircraft in flight (due to dislocated center of stability of the aircraft as result of its advanced design).

reputation was huge. Boeing's CEO resigned few months later. In addition, the NTSB (National Transportation Safety Board) investigation showed that FAA was also responsible for accepting and not analyzing the documentation for initial airworthiness of the aircraft. In simple words: The "green light" for operation of the aircraft was given to Boeing without thorough analysis of those data submitted to them.

This is not the only example how companies decrease their costs of "do business" in the aviation. Another example of cutting the costs is that they do not employ real experts as Safety Mangers or, at least, they are mostly bureaucratic about the nature of implemented SMS. And this is coming from the ignorance of their Top Managers regarding SMS.

6.4.2 Defense Lines of the Regulator

The power of the Regulator[2] comes from its legal responsibility and accountability to oversee the company's performances in the Risky Industries. The oversight activity can be conducted through few mechanisms: Constant or random monitoring, periodic or random audits, voluntary and mandatory occurrence reporting, investigations, inspections, surveys, studies, etc. All of them (individually or combined) can be used by the Regulators to get assurance that companies' performances do not endanger humans, public, assets, or environment.

If the Regulators notice that there are too many incidents in some company, it is a harbinger that the QMS and/or SMS in this company are not fulfilling its preventive role. If the Regulators do not raise oversight audit for the company, they could be treated as "accomplice in the crime". If you are wandering why I am using this expression, please do not forget that when the incident or accident happens in the Risky Industry, the investigation is done by the independent body and the Regulator activities are also investigated in the scope of the ongoing investigation. The simple example for this is FAA and NTSB in USA.

The FAA is a Regulatory Body in USA regarding aviation and the NTSB is (as it is written in their website):

> *... an independent Federal agency charged by Congress with investigating every civil aviation accident in the United States and significant accidents in other modes of transportation – railroad, highway, marine and pipeline.*

The simple example of the independence of these bodies can be emphasized by the accidents of two crashes of new Boeing 737 MAX aircrafts (mentioned in previous paragraphs). After the accidents of the second aircraft in 2019, it produced worldwide grounding of these aircrafts. The investigation of the NTSB was spread also to the FAA and it showed that the FAA neglected their duties during the process of initial airworthiness of the aircraft.

To be more precise, I would also mention the accident with Deepwater Horizon rig (owned and operated by British Petroleum) disaster in the Gulf of Mexico which happened in April 2010. At that time, the USA agency (the Regulator) of

[2] When I write "Regulator" in this book, its meaning is a body established by the State (ministry, agency, etc.) or internationally (organization) to take care for passing and oversighting the regulation in particular area of industry in the State or internationally.

the Department of the Interior that managed the nation's natural gas, oil, and other mineral resources was Mineral Management Service[3] (MMS). Nevertheless, they conducted a lot of audits on the British Petroleum rigs, and they did not notice any wrongdoings (harbingers) on these rigs. The investigation after the accident showed that the archived audit data pointed to BP as a company which had a record of numerous findings regarding the safety of their operations in the past. That was the reason, after the Deepwater Horizon accident, the MMS to be accused for poor Regulatory oversights and as a result, this agency was split into three different agencies.

Having in mind all things explained above, we can treat the incidents as harbingers for the GRe in the Risky Industries. Actually, to emphasize this, I can say that the main point with the GRe in the Risky Industries is misunderstanding that these harbingers (incidents) are predecessors of the future accidents.

There is another aspect of the role of the Regulators in the handling of the GRe. This aspect is not connected only with individual companies but with the industry as whole. Sometimes, the harbingers about the GRe show up at the macro-level, which means these harbingers are global (spread worldwide). The simple example of such harbingers is the financial crisis which started in USA in 2007. As Ms. Wucker explains in her book, there were a lot of signs that things will "explode", but they were continuously underestimated by the US Federal Reserve and other financial institutions in the USA. The main point is that everything started in 2007 in USA, but in 2008 it had already spread all around the world... Other countries were affected also by this crisis.

The companies must take care only for themselves, but the situation with the Regulator is different: It must have in mind the "big picture" (industry itself) as much as "the small picture" (companies).

What can the Regulators do to prevent GRe?

A good Regulator takes care of both aspects of their roles. The problem is that the Regulator must balance between the industry (as a whole) and the companies (as "individuals") on a daily basis. Of course, the first aspect is more important!

It brings us to the very important statement: The Regulator, in the cases where industry is endangered, must give up such individual companies. This is similar to surgeon getting rid of the "bad" organs in our bodies to save the rest of the body. Such a thing happened in 2008 when USA Government did not help Lehman Brothers and they went into bankruptcy. Helping them was simply unsustainable...

6.5 INTENTIONAL OR UNINTENTIONAL GRe

It is very much important to understand that the GRe can be intentionally or unintentionally neglected. If unintentionally neglected, this could be a problem of Human Factors, but also, the reason could be lack of knowledge, skills, experience, or attitude of the employees in the company's daily activities.

The intentionally neglected GRe could happen due to some economic, financial, public relations, or political reasons. Here I would like to present one simple example for intentional "trimming" of data in the nuclear industry.

[3] Actually, after the Deepwater Horizon accident which happened in April, 2010, this agency was accused for poor regulatory oversights and later the agency was split into three agencies.

In Section 2.8 (The Equipment and the BSe in the Risky Industries), I have provided information about the MTBF of the nuclear reactors which I calculated by myself using data from Internet. But there is one beautiful article which can be found on Internet.[4] The article title is "Fukushima, Flawed Epistemology, and Black-Swan Events" and the author is Dr. Kristin Shrader-Frechette from Department of Biological Sciences and Department of Philosophy, University of Notre Dame, Indiana, USA.

In this article, you can find data which explain that situation with the risk from the nuclear industry is not such as presented in the reports of the State's and international Regulatory Bodies regarding safety of the nuclear energy. There are few examples of official reports about trimming the data regarding faults of equipment in the nuclear power plants worldwide. This trimming cannot be assumed to be lying, because there some of the data is "masked" scientifically, just with the intention to undermine the severity of the faults. The trimming does not apply only to the reports regarding faults, but it is present even in the reports regarding the internal audits and investigations triggered after each fault.

All these things, in the article, are described as "political-economic incentives for nuclear accidents cover-up". There you can find the projection of the US Government that all 104 USA nuclear reactors could experience only 1 meltdown accident in 1000 years.[5] This projection is much better than my calculation of MTBF in Section 2.8 (Equipment and BSe in Risky Industries). But in reality, the US reactors already have experienced five meltdowns in just 50 years.[6] In the article, you can find a list of many incidents and accidents which happened in the nuclear industry in the USA and worldwide.

Although, the article (as can be seen by the title) is about the BSe, I would not classify these events as Black Swans. I would like to point here to the GRe, simply because we are speaking about events which are harbingers of accident. We have, in the nuclear industry, a situation where plenty of incidents are just "reclassified" as something with less risk. One of the frequently using methods to do these trimmings was the use of different methods of statistical analysis and different "types" of probabilities to present the events. The word "types" is in quotation marks, because there was no clear definition of the meaning of each of these probabilities used in the reports. There was use of names as "classical probability", "relative frequency," or "subjective probability". Of course, "subjective probability" can provide best results for trimming because it is "subjective" e.g., depends on the person, not on the nature of the methods.

In this article, there is a hidden message that due to pressure of the nuclear lobbyists, intentionally, the Regulatory Bodies tried to present the situation as not so critical, but I would not comment on this.

Instead, what is more dangerous is the fact that many of these reports are produced by professors in established universities. In all these cases, it is clear that

---

[4] Link to the article is: https://www3.nd.edu/~kshrader/pubs/black-swan-2011.pdf (last time accessed on 11th of January, 2021)

[5] In the same article you can find projection of US Government that in all 442 nuclear reactors worldwide, the meltdown would happen once in 250 years.

[6] In the same time, the worldwide reactors had already 26 meltdowns in the same 50 years.

there is intention to produce reports in favor of nuclear industry. Maybe the fact that all these reports from university's professors were made using Government's money could help you to understand why this happened. It is simply unbelievable that all these professors did not make difference between these probabilities and their use…

OK, to be precise: Yes, maybe some of them make the difference, but they used overstated confidence intervals. For me, this is, very much, a strange behavior in area of Risky Industry as it is nuclear industry.

The real question is how we can classify this situation: As intentional and unintentional neglect?

I would not know what to say…

But what I would comment is that I had chance to work in aviation for almost 25 years and I had a chance to meet extremely knowledgeable persons and extremely ignorant persons, not only in areas of safety, but even in the area of engineering. What was scaring me was the fact that for each one knowledgeable person which I have met, I met five other ignorant persons. In the companies, having ignorant persons in some areas is maybe common, but having such a person in the Regulatory Bodies… This is terrible!

So, the GRe are very much connected by the incompetent persons in the Risky Industries and in their Regulatory Bodies.

6.6 CORONAVIRUS (COVID-19) IN 2020…

At the time when I was finishing this book, the COVID-19 pandemic was starting to evolve, and this was a beautiful example of GRe. In the past, many scientists had warned that there is very high probability for humanity to be endangered by a new or novel virus (due to their mutation), but no one cared about it. The SARS epidemic in 2002 was a warning signal also. In 2015, even Bill Gates warned that we need to put more money into the system to stop a pandemic, but again, no one cared. When I said no one cared, I think of politicians who are responsible for funding the scientific laboratories.

When the virus spread all around the world, people started to develop critical health conditions in the form of pneumonia and a lot of them died. Simply, the hospitals were not prepared for such a large number of patients and had no resources to deal with them.

During that time, people stayed at home and worked from there. Social media was the only "window" to the world. I remember that I read a post on Facebook:

> *You pay to the football players and musicians millions of dollars as salaries and only few thousands of dollars to medical doctors and nurses, and now, go to the football players and musicians, do not blame the medical staff!*

Maybe this Facebook post explains the true nature of the humans and the GRe…

I do believe that by the time when this book will reach the readers, the COVID-19 pandemic will not be finished, so it is futile to provide more explanations here…

7 The Invisible Gorilla

7.1 INTRODUCTION

The Invisible Gorilla (IG) is a metaphor and was for the first time introduced in the book *The Invisible Gorilla* (*And Other Ways Our Intuitions Deceive Us*) written by Dr. Christopher Chabris and Dr. Daniel Simons[1] and issued by Crown (New York) in 2010.

It was actually a book about psychological experiment produced and conducted by Chabris and Simons in 1998. They created a video experiment on the students of Psychology at Harvard University, and this later became one of the best-known experiments in the area of psychology. The reason for that is that the experiment reveals, in a comic way, the things which we do not see although they are present in front of us.

7.2 THE INVISIBLE GORILLA EXPERIMENT

In 1998, Chabris and Simons, together with their students, made a short video about two teams of players moving randomly from one side to another and passing between themselves basketballs. One team wore white shirts and the other wore black shirts.

The video was short (a little bit less than 1 minute) and it was copied to videotapes. Their students used the video tapes to show the video all around the Harvard University campus to the student from other faculties. The experiment was to ask the volunteers to count the number of aerial passes made by the players with the white shirts only and, at the same time, to ignore any passes by the players with black shirts.

The point is that, it doesn't matter how many passes were executed because, at the middle of the video, a female student, masked as a gorilla, walked into the scene, stopped in the middle of the players, faced the camera, thumped her chest (as gorillas do), and then left the scene. All these happened in total for nine seconds on the video. After the video finished the volunteers were asked about the numbers of passes and later, the question did they notice something unusual was asked. Approximately half of them stated that they have not noticed the gorilla!

Nevertheless, this experiment was conducted many times with different type of volunteers with different social, cultural, and educational backgrounds, always, approximately half of them, did say that they did not notice the gorilla which was nine seconds on the screen.

The scientific explanation of this phenomenon of how the gorilla became invisible is called "inattentional blindness". This is a phenomenon where the "blindness" is not

[1] Dr. Christopher F. Chabris and Dr. Daniel Simons are prominent American experimental psychologists. They were awarded by the "Ig Nobel Prize" for their Invisible Gorilla experiment. (Ig Nobel Prize is satirical prize for scientific achievement which "first make you laugh and then make you think.")

DOI: 10.1201/9781003230298-7

caused by any kind of damage to the eyes in the humans, but it is caused by the mind. Actually, it is caused by the attention directed (focused) to something else close to the event which cannot be noticed. In the mentioned experiment, the volunteers were so focused on the movements and ball-passing of the team with the white shirts, that they were "blind" to anything else (gorilla!). They had looked at the gorilla by their eyes, but their mind had not seen it. The "inattentional blindness", actually, creates illusions that the things which are, obviously, out of our focus, do not exist.

Strong focusing on something and neglecting other things around is also known as "tunnel vision" in psychology. Such a phenomenon in the economy and in the Stock Exchange areas is known under the name "scarce attention". Somewhere in the literature, you will also find the name "change blindness" associated with the fact that change in the scene (reality in front of us) was changed, but it was not noticed. In this book, I will use the original name for this phenomenon used by Chabris and Simons and that is "inattentional blindness".

To add more "fuel to the fire", the abovementioned experiment was important due to one more thing: A surprise which was showed by all volunteers when they realize on the second watching of the video (this time without counting) that the gorilla was there. Most of them stated that it is impossible that they have missed the gorilla, so their explanation was that the gorilla was included in the video later.

In general, the "inattentional blindness" is a hidden characteristic of humans: The people are capable, so strongly, to devote their attention to something, so they cannot notice anything else in the vicinity of their focus, nevertheless there are also objects or events which can be part of the overall picture. It is important to understand that turning our attention to different details does not mean that we can see everything. Simply, the seeing is not a matter of physiology of the eyes, but rather a process of how the information from the eyes is processed by our brain. Of course, as a cognitive process, it can be very much, affected by our experience.

As mentioned before, the IG is something which is intrinsic in the humans and, as such, it should be investigated by Human Factors in the Risky Industries. Further research of Chabris and Simons showed that there are six illusions which are characteristic for the humans:

1. illusions of attention (invisible gorilla);
2. illusion of memory;
3. illusion of confidence;
4. illusion of knowledge;
5. illusion of cause; and
6. illusion of potential.

Having in mind "the context of the things" of this book, the IG can be associated to many of the "safety animals" explained here. Regarding the GRe, it can be connected by harbingers; regarding the Black Elephant (see Section 8.3 later in this book), a question arises how you cannot notice the elephant. Anyway, in general, they apply to any of the "safety animals". It could not only be associated to NsNhNsM (see Section 8.4 later in this book), because the "No see" part is when humans intentionally pretend that they do not see the things. Someone, maybe, could connect the IG with the OiS also (see Section 8.5 later in this book), but even there, the metaphors do not agree…

Analyzing the IG experiment, the experimenters tried to do the same experiment with changeable subjects and graphically by computer simulation. The experiment was done by replacing the players with the white and black shirts with the white and black letters moving from one to another side of the screen. Here, the point was to count the white letters which were touching the sides of the windows at the same time ignoring the black letters. The gorilla was presented by a red cross which was also moving on the display for nine seconds. Even this version of the experiment[2] showed that, nevertheless, the red cross was highly noticeable and, very much, distinctive from the black and white letters, and almost 30% of the subjects included in the experiment did not notice it.

The IG experiment can explain why the talking on the mobile during driving a car is prohibited by law. Although the driver keeps his eyes on the road, his attention is focused on the mobile conversation. Even if they notice something unusual, their reaction time will be slow. In such a situation, even if both hands of the drivers are on the steering wheel, the reaction will be delayed.

The authors of the IG experiment conducted more other similar experiments providing more complex tasks instead of only counting the passing of white shirt's players. The increased complexity actually increased the unnoticing of the gorilla for additional 20%: Now 70% of the volunteers could not notice the gorilla.

The Invisible Gorilla effect is a situation when you "look, but you do not see". It is a situation exploited by magicians in their shows. Actually, they use the "inattentional blindness" to produce their illusions.

This experiment triggers other experiments that actually showed that "inattentional blindness" does not happen only to ordinary people, but also to experts. There was a study conducted by T. Drew, M. L. H. Vo, and J.M. Wolfe, and the report[3] could be found on the Sage Publishing website. There, it is explained that 24 experienced radiologists were presented with images of computed tomography (CT) of lungs to detect lung-nodule. In the last of the images, the sketch of gorilla 48 times bigger than an average lung nodule size was inserted and 20 of the radiologists failed to notice it. The reason was: They were so focused on searching for lung-nodules that the sketch of the gorilla was not expected to be seen there.

7.3 ILLUSIONS OF IG

The ignorance was mentioned and elaborated in detail in Section 4.4 (The BSe and Ignorance), but here, it can be mentioned also from "the context of the things" of the IG. The point is that, actually, all these illusions affect our behavior, knowledge, skills, experience, and attitude by providing a distorted picture about our private and professional lives and, as such, they can produce errors. Sometimes, these

---

[2] This experiment was conducted later by some of the PhD students of Chabris and Simons and it is known as "red gorilla" experiment.

[3] "The Invisible Gorilla Strikes Again: Sustained Inattentional Blindness in Expert Observers", issued in Psychological Science, Sage Publishing (DOI 10.1177/0956797613479386).

errors can be very dangerous, simply because they are a result of our ignorance about ourselves…

The main point is that, although there are six illusions, it is wrong to treat them individually. They are very much intertwined and, as such, one illusion (if present) could contribute to other illusions. So, let me explain them here individually, and I do believe that the Reader will understand their connections and interactions by themselves.

7.3.1 Illusions of Attention

This is actually the real Invisible Gorilla!

If you cannot notice it, it is because it will not be registered in your mind and it will not contribute to your knowledge. The point is that, the people are prone naturally to this illusion of attention. We do believe that our attention is widely spread to our environment, but it is actually not the case. Whatever and whenever we think that we pay attention to something, we are aware only of a small portion of our neighborhood. We cannot focus on many things at the same time with the same attention. So, for those who say that they are good in multitasking, they are just average in multitasking.

You are shocked?

No need!

The IG gorilla creators conducted another experiment with the same white and black shirt volunteers, but they gave the observers an additional task. Now they should pay attention to the count the number of aerial passes made by the players and at same time they should count the number of the ball passes by the bounces from the ground from one player to another player. Of course, all of it applied only to the team with the white shirts.

The result was this: More than 20% of the volunteers did not notice the gorilla!

Further investigations on the effect of "inattentional blindness" showed that the presented level of attention is mostly individual and it has nothing to do with people's Intelligence Quotient (IQ). The authors did the same experiment on the elite group of Harvard's students and the results were the same. In addition, there was no difference in noticing the gorilla between the men and women.

This illusion of attention is very much present in the traffic incidents and accidents. most of the involved persons in these events usually stated: I was driving normally and looking in front of me, but the pedestrian just popped up in front of me…

7.3.2 Illusion of Memory

The illusion of memory is similar to the illusion of attention: We think that what happened previously is recorded in our memory as it happened. But things are not like that…

Simply, our brain does not spend so much energy to record everything which we suppose we have seen. It is actually a fact that the humans are prone to fabricate things, based on their previous memories about similar events which happened in the past. Actually, they remember the outline of the events or the places and later, their brains "paint" the environmental details based on their previous experiences. This is actually connected with the "narrative fallacy" which NNT has mentioned in his book and I have explained in Section 2.1 (Introduction). Resembling to the illusion of attention, where the people pay attention to the things which they expect to see, the people remember the things which they expect to remember in the illusion of memory.

The reason for that is: Our memory strives to remember the meanings of the things (which, based on our life experiences, are easy to extract), then it tries to remember the details. For example, everybody could remember that he was in the restaurant with his friends yesterday, but he cannot remember the color of their clothes or type of their shoes. Of course, if you ask him about these things, he will try to explain everything, but it will be fabrication based on some other events ("narrative fallacy"!).

This aspect of the illusion of the memory is known also as "change blindness". It means that we are prone to remember that we see our colleagues in our offices, but we are blind to the changes of their clothes and shoes which happen almost every day. Simply, it is not important to us and we fail to notice these changes. Also, it is important to mention that, with time, the illusion of memory increases. This is a situation where our memory fades and our brains try to fabricate the things about the events which happened a long time ago.

One very interesting representation to the illusion of memory is a phenomenon called "déjà vu". "Déjà vu" is a French expression for "already seen," and it deals with false memory regarding something which we experience in the moment but have a feeling that the same event already happened to us in the past.

From "the context of the things" in the Risky Industries, the illusion of memory very much applies to the investigations when incidents/accidents happen. The witnesses of incidents are mostly shocked with the incident/accident and they actually mix the illusion of attention and illusion of memory. That is the reason that it is very much important to record the statements from the witnesses as soon as possible and later, after week or two, to speak again with them. During these (made with time difference) two recordings, it is strongly recommended not to stop the witness when speaking and not to ask any questions when he/she speaks. The difference between these two recordings will provide additional data as regards the integrity of the witness and the trust which can be put in the descriptions of the event.

7.3.3 Illusion of Confidence

The illusion of confidence is very much present all around as and, as such, it can be easily noticed. This is actually a belief that our capabilities to do something are better than in the reality. It can be presented in two ways:

a. To overestimate our capabilities individually or relative to the other people; and
b. To lose or overestimate our confidence to other people.

The first case has a distinctive nature. Usually, the ignorant people are more prone to overconfidence. The ignorant people usually miss the knowledge and the information important to build their self-confidence is based on these missing things. That is the reason that they can easier fabricate overestimation of their capabilities in their mind.

The second case happens when we go to the doctor for some problem and the doctor is using wordings and expressions which give us confidence. Not all the doctors are champions, but their behavior can easily create overconfidence to us, especially when we are afraid about the outcome of the treatment. The overconfidence in the doctors, that they will heal you, is good for the patients, but anyway, it is just an illusion: It will improve your healing, of course! But it will not heal you!

In the case of the overconfidence in the doctors, the biggest problem is changing the competence with confidence. It is not necessary for the patient to build the overconfidence in the doctor, but the patient is not capable to examine the competence of the doctor. So, the overconfidence of the patients in the doctors is not based on doctor's competence, but on their behavior. This is a good example regarding the deficiency of the illusion of confidence.

The illusion of confidence can be found also in sports. Nevertheless, the sport has a competition character where you measure your capabilities on the sport field, still many of the sportsmen are overconfident about their capabilities and it creates wrong expectations during their matches.

An interesting thing regarding the illusion of confidence is that, at the beginning, when the people try to learn some new skill, their confidence is disproportionately high. But as they progress and become better, their confidence is going to became more moderate. The rule of the thumb is: The lower knowledge or the lower capabilities will create bigger illusion of confidence.

In the scope of what was said above, there is good scientific explanation of illusion of confidence and how it is applicable to the humans: It is known as Dunning–Kruger effect.

This is an effect that is experienced by individuals who have assured themselves that their knowledge, skills, or capabilities are bigger than that in reality. It creates a feeling of superiority and arrogance, which is very much associated with the managers, politicians, and people in power. I had many managers in my professional life and I have noticed that every one of them stopped learning when they become managers. Obviously, their appointment as managers was sign for them that the top of their career was reached and they do not need to read anymore: They were confident that they know everything!

When I started to work as Engineering Training Manager with GAL ANS, my job was to provide the training for engineering staff in Air Traffic Navigation Provider (ANSP) in Abu Dhabi. Some of the trainings were Regulatory requirements and some of them were operational requirements. Anyway, no manager attended any of these trainings, which caused me problems later when I tried to explain them what is required by the regulation.

This is a very dangerous illusion, especially for the managers in the Risky Industries. The problem was even reinforced by the Regulatory Body's employees, whose job is to oversee the operations of the companies in the Risky Industry. In aviation, the safety-related positions are subject of Post-Holders. These are managerial positions in the companies which need to be approved by the Regulatory Bodies. The company will send proposal for a candidate for particular position with supporting documentation to the Regulatory Body and if the Regulatory Body's representatives find it satisfactory, they will organize an interview and a short training for the candidates. Anyway, this process is too formal and without any clear criteria.

The same applies to the employees in other Regulatory Bodies. My third book was inspired by one regulatory auditor who was so obsessed with Dunning–Kruger effect that I needed to write a book with the hope that it would change the overconfidence in the auditors.

7.3.4 Illusion of Knowledge

The illusion of knowledge is very interesting!

Actually, whatever we do in our lives is based on our knowledge associated by the information which we have at that time. Usually, we use these associations (knowledge + information) in our lives to predict (or better to say: To guess) what will happen next.

You will say: What is wrong with that?

Well... Using your "knowledge + information" to predict something means that you need to guess the future events and all of this has a value only if your knowledge or information does not change in time.

Another question is: Is your knowledge or available information enough to make such a prediction?

The answer is: Sometimes yes, sometimes not!

We build our knowledge from the available information offered to us. I remember that when my sisters were born (late 1960s), there was an advice from the doctors to keep the infants strongly wrapped as if in a cocoon. When my children were born (1990s), the doctor's advice to keep their arms and legs free...

As you can notice, the things changed in time, but the main point is that I cannot see any problem with both: My sisters (cocoons) and my children (free arms and legs) look to me perfectly normal. Obviously, there is no effect or, at least, I cannot notice it. What is more important, even the medical doctors could not notice it. There are no reports that whatever in this case you have done with your children, they are still healthy or if they are not, it cannot be connected to any of these situations.

The similar changes in the illusion of the medical knowledge could apply to breast-feeding of infants. In the late 1970s where first artificial milks for infants were produced, many doctors (maybe under the pressure of the food companies) advised mothers to move from breastfeeding to these, so called "artificial feedings". Twenty years later, the doctors strongly recommended the breastfeeding by mothers, at least for the first six months.

There are plenty of such examples for the illusion of knowledge. Today, in the industry, you can notice that different experts have totally different opinions about the same problems. It does not mean that you do not need to consult the experts, but the illusion of knowledge obviously will not guarantee success even with the experts. There is a very good book "Eyes Wide Open" (2013) from Ms. Noreena Hertz[4] where you can find data that approximately 17% of the medical doctors make mistakes with

[4] Ms. Noreena Hertz is an English academic, economist, author and TV host.

their diagnosis. In the area of medicine, there is another beautiful book named *Bad Science* from Ben Goldacre.[5] This book is something which must be read by any manager, especially in the Risky Industries. This is a book where many illusions about the scientific results are beautifully explained and these illusions can contribute also to the events in any Risky Industry!

There is another aspect of the illusion of knowledge which is mostly individual. We, as humans, are prone to overestimate ourselves in many areas. Imagine a person who is looking through a window, but he has never been outside the room. He can see the street and cars on the street, people around walking, buildings, houses, trees, and mountains faraway below the horizon. His knowledge is based on the information which is presented to him visually.

OK, he can read the books and watch the TV, listen to radio, go to Internet, but all these sources of information would provide "skewed" knowledge to the person. He could only assume what is behind the horizon seen from the window, but it has nothing to do with the reality. And this is actually source of our illusion of knowledge: We use available pieces of information to fabricate our knowledge and we think that it is enough to make us experts. Obviously, we are wrong…

But what about the illusions of knowledge in the scientific community. I have attended many scientific conferences, and I was very disappointed by the illusion of knowledge presented there. There were cases where a very respected professor tried to introduce word "technoscience" for the word "engineering"; there was a professor from Massachusetts Institute of Technology (MIT) who did not understand that Quality Control will not improve quality; there was a professor who did not know that data could have integrity, but there is no reliability of data; etc. This is actually the problem which NNT meets in his efforts to discuss with arguments about the ignorance about "Fat-tails' and Normal (Gaussian) distributions. In this area, I strongly support him. All these scientists have simply forgotten (or never learned) that our knowledge is surrounded by our ignorance.

There is one more thing which needs to be mentioned here…

I mentioned just a few simple examples, but in reality, and especially, in the Risky Industries, we deal with complex systems, where their complexity of the structure, the processes, and the operations inside could very much increase our illusion of knowledge regarding these systems. So, for those who work there, ask yourself: Are you sure that you know all these systems in details to deal with any situations which could happen to them?

Regarding the complex systems, we use modeling to learn about them or to predict what will happen if some parameter of the system changes. But the modelling itself is also part of the illusion of knowledge. Two scientists will produce two different models of the same system based on their understanding of the system. To be funnier, these models could differ considerably.

The illusion of knowledge is very much connected to the first part of this book where I dealt with the probabilities. I have shown there that we use probabilities in

---

[5] Dr. Ben Goldacre is an author, broadcaster, medical doctor and academic who is dedicated to the research about misuse of science and statistics by journalists, so called scientist, politicians and drug companies. "Bad Science" was his first book.

the cases where we miss knowledge about the things which we would like to predict. Be careful: The knowledge about how the probabilities works could help to fight the illusion of knowledge!

7.3.5 Illusion of Cause

The illusion of cause is something which could very much affect the Risky Industries simply because, there, the imperative is to find the root cause of the incidents and accidents. Establishing a wrong root cause could be very dangerous because it will actually not stop future happenings of the same adverse events.

But, let's be honest: It is not so bad simply because it is very much connected by poor knowledge and poor understanding of causality between two or more events. And in the Risky Industries, this does happen very rarely.

There are a few things that contribute to the illusion of cause.

The first one is already mentioned above: Poor knowledge and poor understanding of the systems, the processes, the operations, or the activities which are subject of your job.

The second one is detecting patterns in the places where they do not exist.[6] This is very much connected with that what was explained in Chapter 1, especially in the paragraphs regarding Determinism, Randomness and Chaos. As you can notice, detecting a pattern is possible only in the areas where we are sure that the Determinism is present. There is no pattern in Randomness, and the pattern of Chaos is full of uncertainty. Anyway, in detecting patterns, our expectations are also very powerful, so if you have some expectations in your trics to understand the behavior of the system under consideration, the detection of pattern (although it is not existing) will be easier.

The third one is determining causality from data where causality actually does not exist.

Surprised that such a thing is existing???

Don't be…

We use very much a probability for the things which we do not know. The probability is based on the data processed by the statistics. Causation means that there is some correlation between the cause and the event, but establishing this causation should not be too bureaucratic. It means, we cannot trust only the data, but we need to relay on the wider picture what is going on. In the science (mathematics), it is known as Bayesian statistics. As explained in Section 2.3.1.2 (Statistics and Probability), in the Bayesian statistics we use data, but we use also some other knowledge associated with particular event under investigation.

In this case, there is a problem also with making difference between the cause and the effect or between the symptoms and the cause. It is pretty much a problem in many observational studies, which most of the medical studies are. My headache is not the cause that makes me feel poor, but the headache is just one of the symptoms because I am ill.

---

[6] The scientific name of this detecting patterns is Pareidolia. For example, we can see faces of people, animals or objects in the shapes of stones, trees or clouds. It is very much present in religion by seeing faces of Jesus or Mother Mary in the shapes or pictures of the nature or things.

The fourth one is connected with the schedule of events. We are prone to believe that the previous events are causing the future events. It could be true, but not necessarily. Do not forget that we live in a complex world and we work in a complex environment. So, not always are the previous events the cause for the next events. Simply, the decision of driving my car is not a reason to the crash which I have experienced later. It is not necessary that something could happen in advance that causes something to happen later.

This is a very interesting aspect of the illusion of cause and you should be careful dealing with it. The causation is, not necessarily, detectable easily, and the schedule of events could sometimes mask the real cause.

7.3.6 Illusion of Potential

The illusion of potential is also common to the people. It means that they assume that some person or something has potential to do something which actually could not happen. Such examples are beliefs that some person has supernatural capabilities to heal you or to change your destiny or that four-leaf clover will bring you happiness. There are plenty such cases where humans assign potential to other humans or to the things, but that potential simply cannot be realized. A widespread belief of such an illusion of potential is so called "sixth sense". The "feeling" about something which is not connected with the human's five senses is simply part of mythology, not of science.[7]

The illusion of potential is maybe one of the oldest illusions embedded in the humans. We can find such illusions in many beliefs of humans, especially in the areas of mythology and religion which were born at the dawn of humanity. Later, it was spread to other aspects of human lives and, today, the abuse of the alternate medicine is the best example.

This illusion can be extremely dangerous having in mind that very often desperate people, who are sick, look for other means of healing. I already mentioned tragic case of Steve Jobs. As it was mentioned in Section 5.2 (The Gray Rhinos), Steve Jobs neglected advice from the doctors and tried to heal himself by some alternate medicines and diets. We all know how it ended…

But there is another side of the illusion of potential which is very much present in our lives or especially in the sports. There, the illusion of potential that you are better than yourself, could boost your performance and you can produce better results. This is known as Pygmalion effect[8] and it was first investigated by experiments conducted by Rosenthal[9] and Jacobsen[10] held in 1968. The Pygmalion[11] effect is a psychological phenomenon where performance of the students can be improved if the teachers have

---

[7] In science, it is known as "subliminal persuasion".

[8] In the literature, you can find also the name Rosenthal effect, because prof. Robert Rosenthal first published paper regarding this effect.

[9] Robert Rosenthal is professor of psychology at the University of California.

[10] Lenore F Jacobson was principal of an elementary school in the South San Francisco. She contacted prof. Rosenthal with proposal to do his research in her school regarding Pygmalion effect.

[11] Pygmalion was (in Greek mythology) a sculptor who fell in love with the statue of woman who was made by him.

higher expectation for the results of their students. Simply, this teacher's expectations are transferred into the students and they push harder.

Further research regarding the Pygmalion effect spread to other areas of lives. In sports, it means that the average team could win against a considerably better opponent if properly motivated by its manager. This is also applicable to the companies where the managers have higher expectations regarding the performance of their employees and, many times, this illusion of potential works.

Obviously, this is a highly beneficial case of the illusion of potential, but do not forget: It is still an illusion!

From the point of the Risky Industries, we should worry about negative (bad) influences of the illusion of potential and at the same time, we should encourage the positive influences of the illusion of potential. Of course, the burden of the use of this illusion stays with the managers who must be aware of good and bad implications. That is the reason that in the collective sports, if the team is losing, the managers should leave.

Underestimation the potential of the Equipment or of the employees, in the Risky Industries, could be very dangerous.

7.4 IG IN THE RISKY INDUSTRIES

Transferring all these previous paragraphs about different illusions in the Risky Industries, the situation worsens. There, as said in the previous paragraph, due to a lot of preventive and corrective measures, the incidents and accidents do not happen very often, which actually affects the expectations of the employees there that they are on the safe side. That is the reason that IG can be very much present there.

Speaking from the point of view of the Risky Industries, most of us have experienced IG driving our cars. Many times, it has happened that a car or pedestrian "popped up in front of us almost from nowhere". The situation (in this case: Illusion of attention) is actually registered long time ago by the adage: Humans see what minds like to see.

Driving our cars, we expect that everyone drives according to the traffic regulation and as such the "popping up of a car, motorcycle or pedestrian" in front of us, is a big surprise. To be more accurate, these types of the adverse events happen most often with the motorcycles which pop-up in front of the cars and usually car drivers are those who did not notice the motorcycle. The pedestrians move slowly and the cars are big, so the not noticing of cars by the pedestrians (and vice versa) does not happen very often. But, the motorcycles[12] are with the right shape and they drive with considerable speed, just enough not to be noticed by the car drivers. Having in mind that there are not so many motorcyclists on the roads compared to the cars and the pedestrians, the car drivers simply do not notice the motorcyclists, mostly because they do not expect to see motorcyclists on the road very often.

To reinforce this conclusion, I would mention the study conducted by Peter Jacobsen (public health consultant from California, USA) in 2000. The study was about the impact of the road traffic on the levels of walking and bicycling in

[12] To the same category belong also the bicyclists.

California. It showed that the possibility a car to hit the pedestrian or bicyclist is lowest in the areas where they are most present. Their presence actually increases the driver's expectations of presence of pedestrians and bicyclists on the road, so the car drivers are more careful.

Transferring this study into the Risky Industry area, we can conclude that there should be the lowest probability to experience any adverse event because the system is made to protect us. If there is a system used to protect us from adverse events, it means the system stops them from happening very often, so expectation to have one is very low. And I can agree with this.

The point is that, when some person is focused to something and did not expect anything else, there is no guarantees that anything outside the focus can be noticed, even if we provide considerable distinctiveness. In the case with "popping up" of the motorcycles in front of the cars, if they wear bright colors, quite different from the environment (phosphorescent or fluorescent), it will increase their noticeability, but it will not provide 100% safety for them. The problem is not the color, but the expectation (or better to say the illusion of expectation) of the drivers!

This "illusion of expectation" is actually a big problem, but I must mention here that our expectations are based on our experience. The experience is based on our previous knowledge and previous events which we have chance to experience during our professional and private lives. In the Risky Industry, having a low expectation means that our experience tells us that it is highly improbable to have an adverse event. But, when the adverse events happen there, our response will be associated by surprise and, as such, it could be non-effective as it has been planned. So, this is actually one of the biggest problems in the Risky Industry: How to maintain focus during operations to provide good response to the adverse events when you do not expect any?

Some would say, we implement emergency procedures, back-ups, and contingency plans associated by preventive and corrective actions. There is, also in advance, emergency and periodical refreshment training how to implement them, but let's be honest: The bad things still happen and responses on them are not always as we expect.

There is one very important study including pilots conducted by the NASA employee Richard Haines and his colleagues Edit Fischer and Toni Price. They used a simulator for an experiment[13] which needed to establish the possibilities of crash of landing aircraft (piloted by pilots with "head-up display") with runway incursion[14] aircraft. The experiment was done in the late 1980s and in that time, the "head-up display"[15] was a new technology available only to the military pilots. The experiment was conducted on the simulator with commercial (civil) pilots who were flying Boeing 727 aircrafts.

---

[13] More details about these experiments can be found on Internet if you look for the paper under the name "Cognitive Issues in Head-Up Display" (NASA Technical Paper 1711). Authors are these three scientists mentioned above.

[14] The runway incursion is a situation when the aircraft taxiing on the aerodrome enters the runway in time and place which is not allowed by the ATCo. It means it must not happen due to possible collision with the landing or taking-off aircraft which occupies the runway in that time! Almost 50% of these runway incursions are pilot's errors.

[15] "Head-up display" is a system (part of the pilot's helmets) which provides many of critical information in a video form, directly on the transparent windshield in front of the pilot's eyes.

It was expected that the commercial pilot will have a benefit from the "head-up displays" in the same manner as military pilots had it, but the experiment showed something else…

During the landing on the simulator, it was noticed that, by focusing on the "head-up display", the commercial pilots could not notice the runway incursion of any aircraft and they could not stop the collision. In addition, it has shown that their reactions to missed approach or other unexpected things encountered during landings were delayed.

The reason why this happened with the commercial pilots[16] and did not happen with the military pilots is hidden in the expectations. The military pilots fly mostly in the military operations where they are trained to expect everything different from the normal and the commercial pilots fly in a highly controlled environment where the bad things (abnormal situations) do not happen very often.

7.5 HOW TO "FIGHT" ILLUSIONS OF INVISIBLE GORILLA?

This is a million-dollar question!

It is so worth because all scientific research told us that humans are not perfect at all and, as such, we are prone to errors. Having in mind that IG is something which is "embedded" into human brains and minds, we cannot deal with surgery of the human brain in this case: The IGs are invisible!

But there are things which we can use to deal with it to make it better…

The first one is by embedding particular experience within the humans. Having in mind that expectations are built on the experience, it means that improving the experience could bring us to more realistic expectations. This is also valid for the life, not only for the Risky Industries. So, dedicating the critical jobs or activities to more experienced employees can really help. Having associated younger employees with their older and more experienced colleagues (during daily activities) will also help to provide more experienced staff in the companies. This is something that needs to be provided by the managers: They must find balance between the "renewal" of the work force in the company and keeping experienced guys there. Encouraging open discussions, later after unexpected things happen, could also contribute to the building of knowledge and experience.

Another important tool comes from the Quality area, and it is known as Poka-Yoke.[17]

For those who are not familiar with Quality, I would like to say that it all started in Japan after World War II and there was an American who started it: William Edwards Deming. He is the father of Quality and his 14 principles how to provide

[16] The similar experiment without "head-up display" and for commercial pilots was conducted by CNRS Research Director, Kevin O'Regan, of the Institute of Paris Descartes de Neurosciences et Cognition which showed similar results. There is book published by Oxford University Press in 2011 and the name of the book is *Why Red Doesn't Sound Like a Bell: Understanding the Feel of Consciousness*. The authors are K. O'Regan and J. Kevin.

[17] The Poka-Yoke is concept first time used by Shigeo Shingo (one of the World leading experts in manufacturing practices in the last century) as part of the Toyota Production System in their manufacturing premises.

good product quality are still applicable today. He was actually the man who triggered the Japanese industrial "boom" in 1960s and 1970s. His ideas, nevertheless neglected in USA, were accepted in Japan and further developed.

Poka-Yoke is a Japanese expression with meaning "error-proofing". It means that you design the product or the service in such a way, so it is impossible to make error by using it.

There are many examples of Poka-Yoke in our everyday lives. The one example of Poka-Yoke is microwave oven. Microwaves[18] can damage human eyes so opening a door on microwave oven when it is in use, can expose the human eyes to these harmful electro-magnetic waves. But building a switch in the door which will cut a power to the microwave generator when the door is open will cease the emission of the microwaves and stop any possible harm to the humans around.

The traffic lights are also Poka-Yoke. There, a particular visual coding (Green for "go", yellow for "be ready," and red for "stop") that helps the drivers to maintain the order in the road traffic.

The wires in the cables for home appliances are also with different colors, so changing the socket or cable will not induce any error, if you know that protective conductor is always green-and-yellow, neutral conductor is black, and line conductor is blue.[19]

In aviation, especially in the Air Traffic Control, there is a particular software embedded into Air Traffic Management system known as Safety Nets.[20] These Safety Nets are actually materialization of Poka-Yoke for the Air Traffic Controllers (ATCo). The software uses the radar and the flight plan data to calculate the future positions of the aircraft during flights and to predict possible conflicts or irregularities in the flights. If any kind of conflict is registered, then it will warn the ATCo by visual and audio signal that something is wrong. The ATCo will have time (approximately two minutes) to resolve the problem. Similar systems, such as Traffic Collision Avoidance System (TCAS), could be found in the aircraft cockpit where the pilots are. TCAS uses data from the secondary radars and, by processing them, it will warn both pilots if their aircraft are on collision course.

Actually, all monitoring and control systems in the Risky Industries are some variances of Poka-Yoke. In addition, there are visual signals (red light blinking with particular intensity) or audio signals (disturbing tone generated to warn us) which can help to control and pay attention to the incoming problems in our operations (processes, activities, etc.)

In general, the fighting of IG is not easy…

As it has been mentioned in the previous paragraphs, there are different illusions and each of them will need a different "remedy". But the point is that, having in mind that they are "intertwined", sometimes solving one of them could deteriorate the situation with others.

---

[18] These are electro-magnetic waves with frequencies from 1 to 100 GHz.

[19] This is color-coding for Europe. In other countries/continents color-coding may differ, but anyway, it is standardized all around the World.

[20] More details about Safety Nets in Air Traffic Control you can find on the link (last time I have accessed it on 10th of March 2021): https://www.skybrary.aero/index.php/Safety_Nets.

7.5.1 How to "Fight" the Illusions of Attention?

There were many trials how we can improve the illusion of attention…

The point is that most of the measures are not permanent and using them the improvement of attention will last only particular period of time. One of the good measures is to have something which is extremely different from the surroundings around you. Having a suit totally covered by red (orange or yellow) fluorescent color for the pedestrians and motorcyclist could help. Actually, this is something which is very much used in the mountains: Most of the clothes of the alpinists, mountaineers, and skiers are of phosphorescent and fluorescent colors with intention, if something happen, they could be easily noticed by the rescuers in the mountains where these colors are not present at all. These colors are not present also in our lives very often, so when you see it, it will obviously make you to pay attention and to investigate what is this.

Regarding the motorcyclist, it could help also very much, but regarding pedestrians, there is a reasonable question: Who will wear red or orange fluorescent colors when going outside?

In the Risky Industries, we have a considerable number of systems dedicated to monitoring and control of the industrial processes. When something goes wrong, the audible and visual alarm will undertake our attention and we could proceed by following the corrective procedure.

The problem with the adverse events (which we do not expect hiding ourselves behind the monitoring controls of our systems) is an IG which we can miss. Having this in mind, we can state that there is no training which can improve our attention. It is more about expectations than about human capabilities or skills. As it had been mentioned before, this is a basic problem: We, humans, are full of expectations. The expectations are based on our experience, and more experienced people have more expectations. Anyway, the unexpected events are called unexpected simply because they are very rare: How many times you have experienced gorillas wandering during a basketball game?

It may seem that the things are not so bad in the Risky Industries: The expertise of employees there could very much help to notice the IG if they happen. This comes from the fact that the same experiments with gorilla were shown to the students who were basketball players and most of them had noticed the gorilla.

Should we rely to that?

Not really…

Let me repeat again: Please note that the IG is connected with the expectations and not by the expertise! As it has been mentioned before: There are many cases in the medicine where examinations done by X-rays or Computer Tomography, where doctors (experts in these areas!) have not noticed the issues simply because they were looking for something else. This could happen also in the maintenance areas during corrective maintenance of the faults. The engineers simply assume what the problem could be and they put their attention on something which distract them to see the real cause of the fault. You can easy notice that what is expected, but you will miss the unexpected, however big it is!

This part of the IG can be easy solved by algorithms of Artificial Intelligence (AI). The implementation of AI into the medicine diagnosis progresses very much these days and the machines are without expectations, so maybe, the future of the SMS in the Risky Industries should be also based on AI.

Another thing which could help with the IG is the speed of operations. Driving slowly could really help with noticing and processing the unexpected events. Slow speed of operation, processes, or activities always give you more time to react and to understand what is going on. Unfortunately, today's life is so dynamic and, trying to achieve everything in short period of time, is a recipe for disaster. At least, today, it is continual habit and not just incidental moments when you are pushing hard and fast.

7.5.2 How to "Fight" the Illusion of Memory?

There is not so much to be done in this area. The humans encounter many things each day during their lives and our brains process this information continuously. Sometimes we fabricate the things, especially when the events happened long time ago.

In the Risky Industry, it is important to understand that there is an illusion of memory, so whenever the problem exists, try to deal with your memory, but reinforce it by consulting operations manuals regarding the processes and the equipment or the Standard Operating Procedures (SOP).

So, do not be in rush trusting your memory: Measure three times, cut only once! It is wise to check before you act!

7.5.3 How to "Fight" the Illusion of Confidence?

The illusion of confidence, especially overconfidence, can be handled by the competition. That is the reason that the sport is good in handling confidence. There are clear criteria in sport, so you cannot neglect them. If you think you are better than others: Fight them! Jump higher or longer, be faster, score more points and goals and this will be a merit for your capabilities.

But how we compete in the Risky Industries...?

The answer is: Through Key Performance Indicators[21] (KPI)!

There is Regulatory requirement in the industry for each implemented management system (and for each company!) to establish KPI. These are usually quantitative criteria for how good your performances in the areas of particular management system, are. For example; For SMS, the KPI can be number of incidents in the last year; for QMS, KPI can be number of scraps in one week (month, year, etc.), for EMS it can be number of incidents in one year, etc. These KPIs can give you good picture about your management system performance.

What is the problem with KPI?

Usually there is strong misunderstanding what can be KPI. Most of the KPI are "cosmetic" or just proforma: They measure something, but the results of measurement

[21] Please note that in the latest edition of the ISO 9001:2015 standard (Quality Management System – Requirements), you cannot find the words: Key Performance Indicators (KPI). It does not mean that you may not to evaluate the performance of your system, but it means that, in the past years it was so much used, so it becomes part of the reality, not of the requirements. In addition, the Quality Manager may establish other methods for evaluation of the quality performance in the company.

of "this something" says nothing. It cannot trigger any action to improve the system and, as such, it is useless.

The well-established and continuously monitored KPIs are valuable measurement of the performance and it can affect company in a good way (realistic confidence), but you should be careful: They can also create an overconfidence. The burden of the KPI stays with the responsible manager, and my experience told me that they mostly fail with the KPIs. To make matters worse, I have been audited (as responsible manager in Quality area) by few professional auditors and even they failed to recognize the real nature and the use of the KPIs.

Producing good and quantitative KPIs, associated with conservative critical analysis, could solve the problem with illusion of confidence!

7.5.4 How to "Fight" the Illusion of Knowledge?

As mentioned in Section 7.3.4 (Illusion of Knowledge), this illusion is also intrinsic within humans and, as such, we cannot do so much in our ordinary lives. But speaking about the Risky Industries, there is one were simple method which will help with this illusion: The 5Why method!

The 5Why[22] method is actually a tool to find a root cause of the problem experienced mostly in the quality area. This is a proven method which pushes the investigators asking by an average of five times "Why it happened?", to find a root cause to the problem. Sometimes, maybe you will need less "Why's" and sometimes, maybe more "Why's", but on average, five times of asking the question "Why", will bring you to the root cause.

If I have headache, I will take an Aspirin (or some other anti-pain medicine), but it will help to deal only with the symptom, which is the pain in my head. But what if this headache is result of tumor in my head? The Aspirin will (maybe) help for two hours, but my root cause (tumor) will still be there and, eventually, it will kill me. So, it is elementary in the Risky Industries not to deal with the symptoms but to try to find the root cause and to deal with it.

Remember, in quality and safety areas in the Risky Industries, during post-incident (post-accident) investigations, we try to find the root causes and with our efforts to eliminate and mitigate the root causes, we provide good knowledge how to stop the adverse events from happening again. The illusion of knowledge would be decreased if we push hard to understand really why this happened, and the 5Why is an excellent method to do that.

The 5Why is extremely simple method, but it needs (again) good knowledge and experience regarding the system from the investigator who is using it. In general, if the 5Why does not provide reasonable root cause, then your illusion of knowledge regarding your system into consideration is still present. Maybe it is time to ask somebody else, who would have different questions and different answers, or at least, a different approach in the search for the root cause.

I would not say here that the 5Why could also apply to our everyday lives. I do believe that it will work, but who would do it on a daily basis?

[22] In some literature you will find it also as 5-Why method.

7.5.5 How to "Fight" the Illusion of Cause?

The illusion of cause in the Risky Industries could be "fought" by 5Why method as well. Whatever is said in the previous paragraph is valid also to this paragraph.

In addition, the thorough investigation, where all possible aspects of the events should be taken into consideration, could also help. This is similar to producing the List of Hazards through brainstorming session. Pay attention to:

- Whichever idea you get, do not neglect it and pay particular attention to it;
- Take care of all possible patterns, but do not be blinded by them. Not everything could be recognizable as a real pattern;
- Take care of the correlations, but investigate different "context" of data;
- Be careful with your inferences and, especially, with your expectations;
- Take care of the schedules of the events, but look at all of them through different angles and aspects; and
- Have in mind that many things could be only everyday habits without any particular meaning.

Anyway, whatever you do, be focused to find the real root cause!

There are many other methods how to determine the root cause, but if you do not implement good and realistic assumptions and do not stick to the data and the facts, they are useless.

7.5.6 How to "Fight" the Illusion of Potential?

The fighting method for the illusion of potential can be expressed by very simple wording: Gather Knowledge!

But its realization, how to gather knowledge, is not easy.

Looking for the explanation how the mythology and the religion were introduced in the human's lives, we do understand that, in the lack of proper explanation of natural or other phenomena, it was easiest way to explain all these events as a Will of the Gods. Today, with the advancement of the science and technology, things have significantly changed…

Better knowledge about the technology used in your systems, processes, operation, and activities could decrease the level of illusion of potential. If this is reinforced by the proper experience, things can become even better.

In addition, if you are a manager, try to determine the capabilities (knowledge, skills, experience, and attitude) to all your employees. Having an illusion of potential about any of them would lead you to dedicate the critical task to an employee who could not finish it.

In Section 4.4 (The BSe and Ignorance), I gave considerable space to explain the influence in the ignorance in treating and explaining the BSe; whatever is said there, it applies also here. Of course, take into consideration "the context of the things" which is different. So, I would not repeat myself regarding influence of ignorance in the adverse events here again.

7.6 EPILOGUE FOR THE INVISIBLE GORILLA…

The Invisible Gorilla is more a metaphor for the reasons or causes for the adverse events than for the real "safety animals". All these illusions discovered by the IG experiment are something which is part of humanity and by trying to fight them, we cannot run away from the fact that we are fighting "humanity". To be honest, the IG is more about human imperfection and its capability to contribute to the adverse events.

There are natural laws and there are human laws. As engineer in electronics, I can speak about natural laws in the electrical engineering. There, the Ohm's law is always the same. It does not depend on the human behavior, moods or feelings, on environmental changes, or on fact that it is raining at the moment. Always, if you connect battery (voltage) to some kind of consumer device (resistance), a particular current will "flow" through the device and the value of this current will be the ratio of the values of the voltage and the resistance.

The human's laws try to mock natural laws. They say that you may not steal, but stealing still happens. So, the human's laws are not always so strict, simply because the human behavior is not permanent: It changes based on our personality and on the Human Factors that contribute to these changes. The humans are prone to emotions and their judgment is mostly affected by their emotions. If I like something desperately and I have poor personality and lack of self-control, the probability that I will steal it is considerably high.

As explained in this chapter, the IG is "embedded" in us, so the IG is actually part of the roots of humanity. Maybe it looks strange, but do not be fooled: Trying to deal with the roots of the IG, we are trying to change humans.

Is it possible…?

I am not sure that I know the answer, but I am afraid that dealing with the roots, we are affecting the quintessence nature of humans and, as such, I am not sure that it can bring more benefit or maybe, it can cause more damage…

So, whatever we do, my humble opinion is that we do not need try to change the humans. I am happy with that how we behave as humans, but we can use something indirectly which will not endanger our roots. Whatever we do, it must help to prevent or minimize the adverse events in our lives. The best example for that is the Poka-Yoke method explained in Section 7.5 (How to "Fight" the Illusions of Invisible Gorilla?), where we really could influence our behavior and our performance, but we still keep our roots as humans untouched.

Additional things which need to be done in the Risky Industry is to put more confidence in the technology and equipment (systems) than in the people. I know that in aviation this is a case with computerized cockpit in the aircraft (Flight Management System or popularly known as Auto-pilot). It does not mean that the humans must be excluded from the Risky Industries. They are still considerably important factor there, simply because equipment does not have experience, which is important factor to realize what is going on. The Artificial Intelligence is going to change that, but we still must rely on the humans. Anyway, there is need to see where and how the humans can be changed by the equipment.

Think about that…

7.6 EPILOGUE FOR THE INVISIBLE GORILLA

[illegible]

8 Other "Safety Animals"

8.1 INTRODUCTION

Working on this book, I had a chance to find a lot of metaphors of different "safety animals" connected by different safety events or safety activities. In the scope of his book, NNT mentioned also the Gray Swan events (GSe) and in plenty of other articles found on Internet, there were mentions of the "Black Elephants", "No seen – No hear – No speak Monkeys", "Ostrich in the Sand", "Red Herring," and "Dragon-Kings" events (DKe). The main point is that all of them are just metaphors used to describe the bad events (or their contributors) in "doing business" or in our lives.

I will try to address all these "safety animals" in this chapter…

8.2 THE GRAY SWAN EVENTS

In his book, NNT has dedicated one chapter to the Gray Swans. In that chapter, the Gray Swan events (GSe) are events which has a huge impact, they are not so unexpected, but they cannot be predicted. Also, the GSe are events which have been experienced in the past which means we know that they can happen. We know them and we can produce models of them to investigate their behavior in different areas. The problem with the prediction is that they can be modelled only as chaotic systems and we do not know when they could happen. As mentioned before, the uncertainty for these chaotic models is huge, and this huge uncertainty is the reason why we cannot predict exactly when they will happen.

Although the GSe are mentioned in NNT's book, there are many authors who are using them as a form of BSe which differs from original BSe, only by the fact that they cannot be predicted. This "cannot be predicted" means that we know that they could happen, but we do not now where and when it will happen. It means that they can be defined as "known unknowns". Speaking statistically, NNT connects the GSe with fractal distributions with "fat tails".

I will try to explain the GSe through tropical cyclones and earthquakes.

We know that tropical cyclones will happen for sure in the tropics at a particular time of the year (known as cyclone season). We know that their impact is huge, but we do not know where and which day they will happen. To be more precise, having in mind that the cyclones are weather phenomena and weather is explained by the Theory of Chaos, theoretically, we can predict them, but prediction could be only valid for next three days and its accuracy will be expressed with only 50% probability that it will happen. There are plenty of historical data regarding the cyclones which are used for modelling, but the prediction with such uncertainty is of no use. As the prediction period (number of days) increases, the error of prediction increases exponentially. Data, statistics, "fat tails," and probability simply cannot help here.

DOI: 10.1201/9781003230298-8

The GSe, with the abovementioned definition, are highly applicable for safety areas because that is what is done by producing the List of (identified) Hazards: We know them, we know that they can happen, but we do not know when and where they will happen.

In his book, NNT has mentioned a possibility to transform the BSe into GSe which could allow us to prepare. The preparation in advance is something which could make impact of the BSe to make them easier, but as NNT stated, not all BSe can be transformed into GSe.

In the scope of the tropical cyclones, this preparation is exactly what happens: When we see the first signs of them (because there are local and global agencies which monitor the weather), the alarm will be risen and humans in particular area will try to prepare for them. Most of the countries in the tropics have contingency plans for cyclones and, mostly, they will execute them when the cyclones approach the land.

In the case of the earthquakes, there is a similar situation, but preparation there happens very long time in advance. There, preparation is mostly done as preventive measures regarding the particular techniques of constructing the buildings. In the areas of considerable seismic activities, there are different regulations how to build seismic-resilient buildings. The resilience of the buildings there should be analyzed during the design phase by modeling and, later, the particular seismic tests for resilience should be executed on the models. It is a "rule of the thumb" that the buildings must be built with structured concrete reinforced by steel.

In the literature, you may find another explanation about the GSe: These are events which have happened in the past and we know them, but their probability to happen again is very low, so they are mostly neglected. In the scope of safety area, the GSe with this definition may not be neglected.

Someone could be confused by the names of Gray Rhino and Gray Swans, but there is considerable difference between these two "safety animals". The Gray Swans are rare (bad) events and we know that they can happen, but due to uncertainty of their modelling we cannot predict them. The Gray Rhinos are (bad) events which can be noticed in advance through their harbingers, but we do not react based on them, which is considerably different from the GSe.

8.3 THE BLACK ELEPHANT EVENTS

The metaphor is made by the combination of two well-known "safety animals": The "Black Swan" and the "Elephant in the Room".

The "Black Elephant event" (BEe) is a metaphor which became popular when it was used by *New York Times* reporter Thomas Friedman in one of his articles, but, for the sake of truth, the term itself, was "innovated" by Adam Sweidan.[1] Sweidan has used the phrase "Black Elephant events" to address the danger of the global warning and ocean pollution with plastics and pollution of water supply on our planet. I would simply add here the higher world population and the danger of food deficiency in the future.

[1] Adam Sweidan is a founding member and Chief Investment Officer of Aurum Fund Management Ltd. and has started charity organization called Synchronicity Earth, dedicated to environment protection.

All of these events are with terrible consequences (they are Black Swans) and everybody is aware about them (as everybody notices the "elephant in the room"), but not one of the politicians likes to disrupt current political balance due to money needed to address them. So, the BEe is a bad event which everyone is aware of and it can provide considerable change of reality, but no one would like to undertake responsibility for action. It is like the fable "The Emperor's New Clothes": Everybody was seeing emperor who is naked, but no one (except one small child) would dare to say that he was actually naked. Having in mind this example, it is very much similar to the GRe and "Ostrich in the Sand" (explained later in this chapter).

From the point of safety and probabilities, the BEe is a bad event which is very probable to happen (which means it can be found in the List of Hazards in the scope of SMS), but it has been marginalized due to some subjective reasons (which has nothing to do with the nature of the event). The BEe could happen in any industry where inertia, denial, or other fear-based and profit-based "drivers", especially within the Top Management, are stronger than the desire and the knowledge to do the real thing which is requested at the moment. It means it could happen also in the Risky Industries and, there, it is more dangerous due to severity of consequences. The BEe can happen very often in the companies which are economically calculative in the area of safety. It means, the Top Management thinks that it is too "expensive" to eliminate or mitigate the possible bad event, so let's neglect it in hope that it will not happen.

As you can notice above, the word "expensive" is in quotation marks which means it is actually not so expensive and it can be handled in the scope of the resources of the SMS, but in the eyes of the company's Top Managers, it looks very expensive. These are companies which usually fake their risk calculations simply because they hardly try not to spend money on safety. The very interesting thing about "expensiveness" of the safety measures is the fact that many companies have not calculated the costs of the bad events if they happen, but they still state that some of the safety measures are too expensive. How they do it with the lack of cost–benefit analyses or comparison of the costs, I have no idea…?

As simple example of the BE today is the situation with the climate change: The experts have warned many times what is going on with the climate change, but the politicians have another opinion by denying it. Of course, the denial is not scientifically supported.

8.4 THE "NO SEEN – NO HEAR – NO SPEAK" MONKEYS

The title of this paragraph is not in accordance with the real one. Actually, this metaphor describes "three wise monkeys" known also as "No see evil – no hear evil – no speak evil monkeys". They are coming from the Japanese tradition (10th century AD) and there is a carving of these monkeys over the door of Tosho-gu shrine in Nikko (Japan). Anyway, they are very much popular in the world as expression of the Japanese wisdom.

The true story comes from the Hindu tradition in India (there are four monkeys originally), which was spread in China by the Buddhist monks and it was accepted by the Buddhists in Japan. The Buddha's teachings said that if the humans do not hear, see, or speak Evil, they shall be spared from Evil. The three monkeys are Mizaru (his eyes covered, meaning "see not"), Kikazaru (his ears covered, meaning "hear not"), and Iwazaru (his mouth covered, meaning "speak not").

Although, the meaning of the wise monkeys in the Eastern Civilization is connected with the Evil, in the Western Civilization, it is a metaphor for not interfering with the reality in a bad way. Actually, the "three monkeys" are often used to explain a lack of moral responsibility in some of the humans who refuse to acknowledge the bad things, looking the other way or acting as ignorant about the subjects. In organized crime (mafia), the three monkeys are a message for "code of silence".

In the (western) business world and in the legal industry especially, there is another meaning: Try to avoid seeing (or hearing) the truth, because truth is sometimes unpleasant. In addition, you may also understand it as: Never speak the truth, if it goes against the cultural expectations of the managers (organization).

From the aspect of safety, these three monkeys have no meaning as events, but they are symbol mostly on the attitude inside the company (especially attitude to safety), where the information submitted by the employees are not taken with due diligence and it is (mostly) neglected. As such, it may contribute, very much, to the adverse events in the Risky Industries.

8.5 THE OSTRICH IN THE SAND

The ostriches, occasionally, when they hide from the predators in their savannas, they stretch out their necks, keeping their heads down, but they never bury their heads into sand. Animal experts, however, will tell you that this could not happen because if an ostrich buries its head in the sand, it would soon die of asphyxiation. Actually, the ostriches use their beaks to turn their eggs several times each day. Having in mind that their nests are on the ground, from a distance, it could seem that as an ostrich leaning into a hole to turn an egg, it could easily look like it has buried its head in the sand.

So, this statement regarding ostriches is actually a myth, but its metaphor (known also as "ostrich effect") is very much spread around the world. It is used in the real life as metaphor for the situations where humans do like to ignore something which is (mostly) bad to them (or to somebody). It is very much used with the politicians in power (mostly dictators) who very often ignore the things which could be damaging for their countries, if it endangers their power. In general, this statement is used for a person who is ignoring the obvious facts or refusing to accept advice, hoping that, simply, by denying the existence of a problem, he will make it go away. In the scope of this definition, we may connect the OiS as cause for the Gray Rhino event. This is actually one of the reasons (and the mechanisms) to produce GRe: Intentionally neglecting the harbingers.

In the scope of the industry (and safety area), this behavior can be used as specimen of managing the companies without addressing the real problems. To be honest, sometimes it happens (unintentionally) to the managers, mostly as a result of the ignorance regarding the causes or consequences of the events. And sometimes it can be done intentionally by these same managers to protect themselves, their companies, or their positions.

The OiS is also very much present when the bad events happen in the companies. Usually, the CEOs of the companies are among the first ones informed. Receiving the information about accident in his company is not easy. During the disaster, at the first moment, they are prone to shock, which could develop into fear and panic later, maybe disbelief and, eventually, during investigation, it will result in denial putting their "head in the sand". The CEOs will hope it goes away, and of course, they will deny their responsibility for it, very often blaming someone else.

In the psychology, it is known as "unintentional blindness". There, the overall metaphor of "Ostrich in the Sand" effect is also known as "selective exposure hypothesis" where humans must make decision how to continue operation based on new information, but they ignore the information depending on the nature of offered information (good or bad).

Anyway, it is very much present in our lives and the safety (as a discipline) cannot neglect such events. Unfortunately, when bad things happen, somebody will always behave as "Ostrich in the Sand".

8.6 THE DRAGON-KING EVENTS

The Dragon-Kings events (DKe) can be defined as events extreme in their impact, but they do not belong to the same population as the other events which are intrinsically part of the system. The idea about DKe was presented by Didier Sornette in 1998, and he is still the leading expert in the DKe.

The name should be considered as a dragon (a mystical animal with extreme power) associated with the king who is "outlier" in the set (population) of the rich people in the country. Here, the "Outlier" has not the same meaning as "outlier" in the statistics. There, the "outlier" needs to be discarded from all population. In the case of DKe, these are outliers in particular set of data, simply because they are coming from another set of data. We can define them as statistical noise and in their own set of data, they cannot be defined as outliers.

The king is "outlier" in the "set of the rich men" in the country, because his richness comes from the "aristocracy", not from the "business". He has not earned the money from the factories which are producing goods, but his money comes from the legacy of his family and from the State.

The DKe appear in highly dynamical systems (operation or process which change in time and/or space) as a result of amplification of small events (parameters outside the system which can be characterized as a noise) that are not fully active during each phase of operation of the system. So, this noise can show up during particular phases of operations, but it is not intrinsically part of the operation. Anyway, there is something in the operations which can "feed" or "amplify" the noise, making it enough big to catastrophically disturb the operation.

To explain how DKe happen I will use the oil prices...

In general, the market price of each product or service is established by the ratio of volume of selling and volume of offering. If the selling of the product is good (customers look for the product), its price will increase. Regarding the offer, it is opposite: If the offer is too big (bigger than request from customers), the price will decrease. But, the price of oil may depend also on other factors. In the case of war (which is not intrinsic to the market) in the countries which are producing oil, the price of the oil on the world market will increase.

This increasing of the oil prices in the case of war or other instability, is actually a Dragon-King event. The DKe pops up when something abnormal happened in the system of oil market and the situation with prices is amplified by the noise (war), so the disturbance on the market is created. Maybe for someone it looks not so strong, but some State's economies could fall in recession due to dependence on the oil import or oil export. And if you would like to say that the oil has nothing to do with wars, then think twice: The wars are fought for some benefit and controlling the oil production is a huge benefit!

In the safety areas, the DKe can be described as avalanche accident during skiing in winter time. The good skiers like to challenge the nature by skiing in the areas which are not marked as safe, so their skiing (or shouts and yelling produced by the skiers) there can easy trigger an avalanche and the accident can happen. In other words: The operation or the process (skiing) disturbs the environmental conditions (amplifying the noise) and the avalanche (DKe) happens.

Additional example for the DKe is a wrong drug given to the patient due to wrong diagnosis. In this case, it is obvious that drug belongs to different set of illnesses and it cannot help with the true illness. In the contrary, it can produce additional medical problems, if used long time and in big quantities.

The DKe can occur also by other factors, not necessarily triggered or amplified by the parameters in the operations. Such an example can happen in the aviation. The sudden change of the weather conditions over a particular area will affect the air traffic in this area to the level of possible accidents. The flight of the aircraft is a dynamic system is highly dependent on the weather conditions over the flight areas. It means that the weather is a "noise" to the flights and the weather changes can be "amplified" by the changing environmental conditions, so the DKe is very much possible also there.

There is article "Rainfall and Dragon-Kings" from O. Peters, K. Christensen, and J.D. Neelin issued by *The European Physical Journal* in 2012 which is supporting the idea[2] that most of the hurricanes are DKe.

The DKe are also very much present in the industry, especially in the dynamic areas where the control over the system is achieved by feedback loop. Delaying feedback signal to the control circuit (due to any reason) will produce delay in the response and the controlled circuit could easily go out of the tolerance ranges. Having in mind that the feedback circuits are very much used in the control and automation of industrial dynamic processes, it is easy to understand the effect of the DKe there.

Having this in mind, we can say that, from the safety point of consideration of the DKe, they cannot be predicted, but to notice them in advance, there is need to spread monitoring not only on parameters of the operations of the system but also to the associated environment and to the possible noises (obstacles) from other sources. In general, the DKe support the theory that each accident happens as a "chain of events" (Swiss Cheese Model) and not necessarily one simple thing could cause the accident.

[2] For the sake of truth, there are some limitations in the article, so the authors limit the support of idea that hurricanes are DKe in accordance to these restrictions. The article says: "at least, the hurricanes are princes, if not kings".

The point of importance with the DKe is that they can happen, but the hindsight analysis later could not always reveal what was the cause of the event (amplified noise or something else caused change of the parameter of operation). In the statistical investigation of the DKe, there are statistical tests which do not focus on the tails of particular probability distributions, but they deal with the cumulative distribution. Using the cumulative distributions, the information hidden in the tails can be found (cumulatively) at the beginning and end of the distribution. It makes the information hard to understand.

Going back to the example with the TTL circuits, explained in Section 2.5 (The Outliers and the Black Swans Events) and Figure 2.8, it is obvious that the shaded area between 0.5 and 2.7 V is the area where the DKe could happen.

The best statistical tests for finding the DKe are Confidence Interval Test, U-test, and DK-test, which belong to generic methodology to check if the tail of one distribution belongs to the probability distribution of the operation itself. The most important point is that all tests need data and, as such, they cannot be used for prediction. So, they can be used only in hindsight, after the investigation, to explain that it happened due to the DKe. Also, they can be used to determine if the event was a DKe, but they cannot determine what was the cause for the DKe. In addition, a particular knowledge and experience in own operations and processes is necessary to critically analyze the DKe to determine the cause.

Most of the research regarding the DKe events is done in area of macro-levels, which I would explain as huge natural disasters such as earthquakes, hurricanes, floods, volcanic eruptions, etc. On the micro-level, which means that these are DKe which can be associated with incidents and accidents in the Risky Industries, there is not enough data for considerable analysis. Anyway, the DKe can be associated with the dynamic systems or in our case with the Theory of Chaos's associated events, which could be part of the operations in the Risky Industries.

The most important think in fighting the DKe is that the employees first need to understand the dynamics of the system in a systemic way and holistically. It means that they need to understand the possibility of existence of the DKe as result of the dynamical evolution of their system.

In general, as it is mentioned above, we may try explain a lot of incidents and accidents in the Risky Industries as DKe, but only in the hindsight!

8.7 THE RED HERRING

The Red Herring (RH) is an expression about something which is used to distract the humans from the true nature of the things. It can be something false (statement, expression, etc.) or it can be something true, but without any connection to the real problem. The RH could be the fake news which will turn the public attention from something which is, very much, in the center of popularity, but it is unpleasant for somebody.

The metaphor "Red Herring" comes from the herring which is dried on a smoke and after drying, it changes color, from a light cream to red . Actually, noticing the red color on the herring means that you can eat it.

The RH, as "safety animal", could be intentional and unintentional.

I am sure that any of the readers can find an example for RH, but the beautiful one, which popped up in my mind, could be found in the Hollywood movie "Wag the Dog" (starring Robert De Niro and Dustin Hoffman). There, the "spin doctor" (Robert De Niro) engages the Hollywood director (Dustin Hoffman) to provide (intentional) "Red Herring" to displace the public attention from the ongoing sex scandal in the White House. The RH is the fictional war in Albania and in the movie, the press and public change their attention to this war, neglecting the sex scandal.

In "the context of the things" in the Risky Industries, the RH could be only unintentional, simply because if it is intentional, it will be part of the Security concerns, not the Safety concerns. The RH could be connected with the GRe (misinterpreting the harbingers) and IG (illusion of attention) and its effect on the operations in any company in the Risky Industry could be devastating.

9 How to Fight "Safety Animals"?

9.1 INTRODUCTION

In the previous chapters of this book, different "safety animals" were explained and in this chapter, I will try to point to some general measures or activities which could be undertaken to eliminate or/and mitigate situations with these "safety animals".

I will mention again the extended equation for probability from Section 2.3.1.1 (Extension of the Definition of Probability):

$$P(\text{adverse event}) = \left(\sum_{n=1}^{k} p_k \right) + P_{\text{NA}} = 1$$

The main importance of the P_{NA} in the equation above is that our present Safety Management System is too much based on known matters than on unknown matters. It means the total approach must be updated or upgraded. This book is trying to help achieve it.

The already presented general plan for evolving of Safety Management Systems in the industry from proactive to predictive is very much endangered. Things, especially "safety animals" considered in this book, tell a story that things are not always simple regarding their predictions and we must put more efforts, with more agility and more innovations, to change the overall perspective in looking at and managing the safety issues.

Some will say that we fight to improve Quality and Safety by establishing particular management systems, and I strongly support this view. My problem is that during my professional life, I have encountered so many bureaucratic management systems which are, actually, a factor to increase the number of incidents and accidents.

Maybe you are shocked by this statement, but I stand behind it!

Having a management system implemented, but not having a knowledge how much effective and efficient it is, could only provide fake self-confidence between the Top Managers. I do not know what you think about it, but unsupported self-confidence (someone call it "arrogance") could be considerable as a factor in contribution to incidents and accidents, not only in the Risky Industries but also in our everyday lives. My message to the Top Managers is expressed by an old Slavic adage: "The fish starts to stink from the head".

Whatever is the "safety animal" which can endanger safety of humans, assets, or environment, the good monitoring system could help. Maybe it will not help to eliminate the "animals" immediately, but it can prevent a lot of adverse events and can deal with the containment of the consequences. Of course, it needs to be supported by particular control system, emergency procedures, and back-up and

DOI: 10.1201/9781003230298-9

contingency plans on how to deal with the operations and consequences if something bad happens.

Whatever happens, the good monitoring system will rise an alarm, and after this, it is of utmost importance, to react fast. The speed of the reaction could change the outcome of the adverse event. The employees in the company must be capable of dealing with these adverse events faster than these events evolve into something more catastrophic. The well-prepared emergency procedures and trained employees will help in such a situation.

Anyway, the things are not so simple, so every Safety Manager must understand the complexity of his own company's (its own "risk landscape"), the established and implemented management system, and the complexity of the activities connected with operating and maintaining this system. The most important thing is to understand "the context of the things" associated with his company and his industry. Everything else is implementation, monitoring, and control.

But it is worth to say here: Whatever you decide to do, being conservative and cautious in your approach of dealing with any of these "safety animals" is better than being radical and brave!

Let's speak about different things which can improve situation with all these "safety animals".

9.2 AREAS AFFECTED BY "SAFETY ANIMALS"

Roughly, there are three areas of risks regarding the well-being in each company (in each industry) and they are more economically related.

The first one is the "financial" area. This is connected with the care of finances in the company, but mostly with the amount of money available to conduct its operation for a considerable period of time. The second one is "operational" area, and this is connected mostly with maintaining normal operations without faults of equipment and failures of operations due to many reasons. The third one is "strategic" area, where there is possibility to fail to adapt the company on the novel market requirements considering amount or quality of the products.

From the point of safety in the Risky Industries, the most critical is the "operational" area. This is the place where all "safety animals" can show up with catastrophic consequences, not only for company but also for the city, state, environment, etc. I said "most critical" because whatever "safety animal" happens in this area, the consequences are caused by the faults and failures in this area.

The other two areas ("financial" and "strategic") can contribute indirectly to the problems in the second area ("operational"). Having financial problems of the company will (maybe) produce decrease (firing) of the employees, which means that those who will stay will have more jobs to do in the same time. This will increase the stress to these guys and, in combination with the fear that company could go into bankruptcy, it will be a considerable Human Factor to deal with. The "strategic" area is also similar by the results which can show up. Failing to adapt to market requirements in timely manner will push the Top Management to produce desperate measures to "compensate" the missing time and missing opportunities. This will produce pressure to the employees and another Human Factor will endanger the company (city, state, environment, etc.).

The main point is that, any "safety animals" can show up in any of these three areas, but directly, the Safety Manager should take care only for the "safety animals" in the "operational" area. Of course, he must monitor the overall situation in the company, but his responsibility is in the "operational" area. Whatever is done in these situations must be a team effort, which means the all managers must analyze implication in all areas and produce compromise of actions and measures. This is a job which is not easy to achieve in the complex and big companies. There, the Top Management must have more sense for understanding the situation and for realistic compromises.

The "strategic" area is very important because the proper safety strategy needs to be implemented also in the company to deal with the future development. The "Operational" area is mostly static during everyday activities in the companies, but "strategic" area is responsible for the "dynamic" future of the company. The change of the strategy will affect the operations in the company and, as any change, it will solve some of the problems, but it will create new and novel ones also.

The strategy is established by the Top Management and the tactics, how to implement the strategy, is part of the duties of the line managers and other employees. But whatever is done, there is no need to say that (in each industry!) the safety issues and dealing with their consequences must have priority!

9.3 TALEB'S "PLATONICITY"

In his book, NNT speaks about "narrative fallacy" and ""platonicity"". I have briefly explained what is "narrative fallacy" in Section 2.1 (Introduction) and here I will focus on "platonicity".

At the end of NNT's book, there is a Glossary, and there is a definition of "platonicity" there:

> *The focus on those pure, well-defined, and easily discernible objects like triangles, or more social notions like friendship or love, at the cost of ignoring those objects of seemingly messier and less tractable structures.*

I would simplify this definition…

The "platonicity" applies to our ordinary lives when we try to explain the unknown things by known things. In such cases, when we encounter something unknown, we try to explain it on the basis of the things which we already know and have encountered previously. It is, very much, similar to "narrative fallacy", but with "narrative fallacy" we create a story, but with the "platonicity" we simplify and change the complexity with simple parts.[1]

It means that we are limited to our knowledge and previous experience, so we miss a chance to find better (and maybe, more appropriate) explanation. It can be called "convergent thinking". Opposite of this is "divergent thinking" where we are

[1] In the science, the "platonicity" can be found as a method called Reductionism. This a method which helps in understanding of complex processes and systems by "reduction" of these processes and systems on their constituent parts. It will help to understand better the complexity, but you need to be careful with the use of Reductionism: Every simplification result in missing information or functionality which could be very important to obtain the clear picture what is this about.

looking for solution outside of the problem (outside the box), which means: We are open for new and novel ideas for the explanations regarding unknown things.

In general, the "platonicity" is a way of thinking, where, as an example, we explain particular complex forms or shapes (systems!) by their simplification in regular forms or shapes (triangles, circles, rectangles, etc.). It can provide explanation, but we cannot run away from the fact that we neglect complexity, which is maybe essential for the explanation of these complex systems.

In the sciences (mathematics), a simple example for the "platonicity" is the linearization in the non-linear dynamics. There, the non-linear phenomena are described by differential and difference equations, and differential equations can be strongly non-linear, which makes them very hard to solve. To find solutions to these "strange" phenomena (especially for those described by differential equations), we use well-known methods of linear mathematics or graphical presentations. But the point is that the linearization does not provide the exact solutions and they are very much limited to the area where we use the linearization. We cannot predict all these phenomena, nevertheless they are deterministic. The Chaos is the highest level of non-linear dynamics and I already mentioned that it is highly unpredictable.

The "platonicity" pushes us towards modelling in the science. The modelling is very powerful tool to examine different things in the science and engineering. Nevertheless, it is very much subjective, because it is based on our understanding of the things (which is pure "platonicity"). In simple words, it means that the same phenomenon can be modelled by two persons in two different ways, depending on their understanding of the phenomenon. In the case of modelling, it is wise to think about risks which are neglected by our model. Do not forget: The safety matters, not the model!

In the scope with this aspect of the "platonicity", I have already mentioned the Chaos. There, it is proven that the Chaos can show in two areas: In time and in space. The Chaos in space is represented by fractals which actually explain that very complex and simple forms can be scaled to infinity. The fractals in the Chaos are in the complex area, so it will not be explained here, but just remind yourself that the uncertainty will exponentially increase with the Chaos.

The "platonicity", in combination with the "narrative fallacy", makes us think that we understand the situation more than we actually understand it. As such, it is not good to have it in the Risky Industries. As NNT explains in his book, "platonicity" is actually reason to use the Normal (Gaussian) probability distribution instead of the "Fat-tails" distributions in the Stock Exchange area.

The "platonicity" is very much a problem especially with the BSe and the GRe. In both cases, the "platonicity" does not allow us to be creative in the recognition of the BSe and the GRe. The adage that "The devil is in the details" is very much neglected with "platonicity": Neglecting the details about those events could affect the determination of the outcome (success or failure) in handling them.

The risk should be considered locally and holistically, but the details must not be neglected. Maybe the full story about the "safety animals" is hidden in them…

Going back to the equation from Section 2.3.1.1 (Extension of the Definition of Probability):

$$P(\text{adverse event}) = \left(\sum_{n=1}^{k} p_k \right) + P_{\text{NA}} = 1$$

I can state that all problems with "platonicity" are hidden in the term P_{NA}.[2] And again: Do not neglect it!

That is the reason that there is need to have particular personality to be a Safety Manager. The person chosen for a Safety Manager must have few important attributes: Knowledge, skills, experience, and attitude. All these attributes, in synergy with open-minded personality, could determine destiny of the "safety animals" in advance.

As I have stated previously, in the P_{NA} are hidden all randomness and uncertainty of any adverse event in our lives and they are very much place for the BSe. So, anyone needs to be careful with the P_{NA}.

Being obsessed with P_{NA} where the BSe are means that it can happen we neglect the GRe or some other of the "safety animals" and, again, the bad things will happen. Do not forget that the GRe are known and they are part of the List of Hazards with clearly determined risks for each of them. So, putting so much emphasis on P_{NA} and neglecting the first part of the equation from above can "provoke the devil to come into your courtyard". That is the reason why our SMS must be holistic.

9.4 THE GRAY RHINO'S HARBINGERS

As it has been mentioned in the Chapter 4, the harbingers are connected with the GRe. For the BSe and the Chaos, they do not exist by definition, so there is no need to think about them in these areas.

But in the areas of the GRe, the harbingers should help us to understand that GRe is coming. The problem with these harbingers is that:

a. They cannot be always noticed;
b. If they are noticed, they could be neglected; or
c. They could be noticed and be analyzed, but (again) be neglected.

Not noticing 'harbingers" of the GRe could make them to show later as the BSe. And this is the worst situation. The impact is huge, and the response is pretty much confusing. The reasons for that are different, but it is dependent on the quality of monitoring the processes. More details regarding "quality of monitoring", you can find in Section 9.8 (Real-Time Monitoring).

The situation that the harbingers have been noticed and have been neglected make them real GRe, and it is interesting to analyze why this happens. My father used to say that "only swimmers can drown". The people who do not swim, they keep themselves far away from the water. And this statement applies to this situation. Neglecting the harbingers can happen only by the guys who feel overconfident in

[2] P_{NA} stands for probability of event which is "Not Assumed" and in the equation is used as synonym for probability of BSe (P_{BS}) to happen.

their knowledge about "what can go wrong". I must be honest with you: I cannot submit many reliable data to support this opinion, but my experience told me that in most of the cases, even the analysis regarding the harbingers is not done, they have been neglected immediately by the people who "know that this is nothing".

The most critical problem is the third one: You notice them, you analyze them, and (again!) you neglect them, because the result of analysis (?) says they need to be neglected. In this case, definitely, something is very much, wrong. Having in mind that the analysis can be done by using qualitative methods (FMEA, or similar) or by quantitative methods (FTA, ETA or similar) the problems, mostly, are not in the results, but in the interpretation of the results. Whatever the analysis is done, the results are also subject of interpretations. I know that in aviation, there are not particular values for frequency and for likelihood (probability) which can help categorization of the risks, so very often it happens that two different analysts use the same data and same methods, but still, the interpretations of the results are quite different. The reasons for this are pretty much complex. "Complex" because there are lots of personal, educational, cultural, and religious factors behind each interpretation.

The simple example for this is the financial crisis in 2008. For NNT, it is a BSe, but having in mind that there were some warning signals from many economists before it happened, it can be treated also as "failed" GRe. "Failed", because the warning voices were neglected by the factors which could change something. The main reason, in this case, could be found in the eternal misunderstanding between the economy and the politics.

In the case of Safety, the similar misunderstanding exists, between the Safety Managers and Top Management. In my humble experience, most of the Top Managers are focused on money saving and every safety warning from their Safety Managers is considered through the prism of saving money. The reason is obvious: They are put in this position to bring money to a company, so they are happy to accept some amount of risk, only if it can help to save/bring money.

To be honest, there is nothing wrong in that if, in hindsight, everything is OK (nothing happens). But the point is that you cannot know how the things will show up in the future. In such cases, it is wise to do Cost-Benefit Analysis (based on ALARP principle[3]) which will consider preventive and corrective measures before and after the event. In addition, it is prudent to be a little bit more conservative during this analysis.

The principle of functioning of the Safety Management is that we prepare ourselves for the bad things which we know that could happen (List of Hazards). For the things which we do not know that they can happen, we try to be resilient and to eliminate and mitigate the consequences. Anyway, the biggest problem is that there is need for a compromise: How much money is enough to be safe? Answer for this question could be known only on hindsight.

In general, there is no 100% method which will advise you what to do. Maybe it is good to trust your instincts (if they were good in the past), but even this one is a subject of uniform probability distribution (chance) which is actually guessing. Do not forget, guessing is full of randomness and uncertainty!

---

[3] Will be explained in the next paragraph.

9.5 DEALING WITH THE CONSEQUENCES INSTEAD OF THE CAUSES…?

Having in mind that speaking about "safety animals", mostly we have a problem with the P_{NA}, it is clear that this probability cannot be maintained in useful manner. There are too many unknowns within the P_{NA} and uncertainty there is very high. So, there is need to find another way that can cover the bad things which are hidden in this type of probability. Obviously, looking here for hazards included in the List of Hazards cannot help.

There is another very important aspect of the adverse events strongly connected by the consequences which can show up. This is the dynamics of how the adverse events will show and how fast they will develop in the company. Today's industrial environment is highly dynamic and, as such, the processes inside the companies must have the ability to adapt to this changing environment fast. The interchange of information (bad or good news) in today's Internet era is also very fast, so it puts pressure on dynamics of happening of the events and responses to these events.

It is logical to realize that the high speed of developing the adverse events could increase the severity of consequences, so the speed of the response should be also very high. The high speed of responding can be only achieved if we have prepared the responses in advance.

For dealing with the P_{NA} and for providing adequate speed of the responses, it is good idea to move from predicting the causes to preparing for the consequences. It does not mean that we need to cancel our present aspect of the SMS, but it means that we need to upgrade it to better prepare for the consequences. As much as we started our SMS activities in the company by producing the List of Hazards, now we should focus to establish also the List of Consequences. This list will help us to deal with the BSe and IG especially.

The establishing a List of Consequences shall be done on the same style as the establishing of the List of Hazards: Through brainstorming sessions. The team which will deal with that must be multifunctional, and choosing the team members will depend on the complexity of the processes inside the company. Same as for the List of Hazards, it will be a multifaceted job where each part of operation will have its own "context of things".

Establishing a List of Consequences is just the first step in dealing with the adverse events hidden in the P_{NA}. The next step will be how to produce preventive and corrective measures which will be used to eliminate or mitigate these consequences. Here, the agility, the creativity, and the innovation must also take place.

The most important thing in dealing with the List of Consequences will be to focus on the phases of the response: There should be two phases.

The first one starts immediately when the adverse event happens: It is to deal immediately with consequences and it is not easy to do that…

In the time when the adverse event will happen, too many things will happen in a very short period of time, so just tracking them will be a problem. In this area, prepared emergency procedures, good back-up strategy, and good contingency planning, associated with good training provided to the employees, can make the difference between life and death.

The second phase is to start immediately when the immediate consequences are maintained (eliminated or mitigated) and there is need to deal with the causes. If there is possibility to merge these two phases, it is excellent: The dealing with the causes should be extremely useful for the quick response measures, because, in addition, resolving the causes will help also with consequences.

The simple example for this is: If the cause of fire in the building is the fire in the gas storage in ground level and fire was spread to higher floors, then firemen should focus to contain fire on the ground floor (where the flammable gas is) instead of on the higher floors. The dealing with the ground floor will help to extinguish also the fires on higher floors. If not doing that means: The fire on the ground floor will continuously "feed" the fires on the higher floors.

But this should is a matter of good planning and it is not easy to be realized due to the nature of the adverse events and uncertainty connected with them. This measure is highly dependent on "the context of the things" for any company and any "safety animal".

Regarding the List of Consequences, there are two aspects of each of the corrective and preventive measures used for elimination/mitigation of the consequences: Effectiveness and efficiency.

Effectiveness can be connected by the merit of achieving effect of the measure (does it work and how good it works). Efficiency can be connected to the costs and resources spent to implement these measures. The effectiveness is of utmost importance because efficiency is mostly an economic category for the companies. It is clear that in the case of the adverse event, the needed resources to cope with it will be bigger than those needed for the normal operations. A wise planning of the resources must be part of preventive and corrective measures as part of the List of Consequences.

Anyway, in the Safety, there is principle known as "As Low As Reasonably Practical" (ALARP), which says that there is no reason to put too many and too expensive resources to maintain safety, because it can put the company into bankruptcy. The wise approach is to focus first to providing as much as possible high effectiveness and then to do a cost–benefit analysis to establish real efficiency of the measures. Of course, it will be done by humans which means that all their imperfection could be built in the measures and in the cost–benefit analysis. Do not forget: The industry is working for profit, and increasing the costs will decrease the profit. Top Managers are very much aware about this, so particular bias on the side of profit in the managerial decisions based on the cost–benefit analysis is very much expected.

9.6 ANTICIPATORY FAILURE DETERMINATION (ADF)

As a method to fight the BSe, very often, you can find in the literature an expression "Think the unthinkable". And this applies to this paragraph. Obviously, there is need to provide a paradigm shift in our approach to deal with all "safety animals," and it could be "unthinkable".

I already have mentioned in the Preface that the term "Black Swan" popped up to my attention when I found the Master's Thesis of Adesanya Adeleke Oluwole which was submitted at Faculty of Science and Technology at University of Stavanger

(Norway) in June 2014. The thesis actually is investigation about the possibility to predict the BSe by the use of Anticipatory Failure Determination (ADF) method which is actually based on Russian TRIZ[4] or in English known as Theory of Solving Inventive Problems.

The AFD is a method based on the procedures to "invent" possible "unknown" failures in a structured but creative way. So, it is not working with common and already known failures (which can be part of the List of Hazards[5]), but it is "innovating" all other possible failures (not included into the List of Hazards). It was produced in the industry as a "remedy" for the FMEA (Failure Mode and Effect Analysis), because FMEA is working mostly with known failures. There are ways how to combine AFD and FMEA, so it is possible to work in parallel with both methods. For the automotive industry, the FMEA[6] is part of standardization (IATF 16949 requirement) and the AFD can be used there as upgrade to the FMEA. Such a "hybridized" AFD-FMEA is called Failure Mode and Effects Anticipation and Analysis (FMEAA).

The AFD can be divided on two parts: The Anticipatory Failure Analysis (AFA), which needs to provide data for the Anticipatory Failure Prediction (AFP). The AFA is a systematic procedure for identifying the root-causes of a failure (or other undesired phenomenon) in the operation carried out by the system. Later, these data are used for making corrections in the system in a timely manner. The AFP is a systematic procedure for identifying beforehand (for the purpose of preventing) all harmful events that might be associated with the operation carried out by the system.

The AFA, actually, requests the safety system in the company to be organized "up-side-down". The meaning of "up-side-down" is that the Safety Manager must try to "attack" the system that he is paid to protect. Each Safety Manager must be familiar with the system in detail, but now, within AFA, this knowledge must be used to find a not known way to make the system to fail. So, instead to be a Protector, the Safety Manager must put himself in the role to be Attacker (Intruder, Terrorist, Saboteur!). Behaving as Attacker, the Safety Manager will have better possibility to "innovate" different type of "attacks" which cannot be considered that they can happen when he is in the role of Protector. In this area, "the context of the things" is very important. It assumes that "the context of the things" of the Attacker is quite different than that of the Protector. Simply, by accepting "the context of the things" of the Attacker (thinking as him), the Protector will be more successful in the protection.

All the possible scenarios for the failure of the system (in the scope of AFA!) should be considered (analyzed). The results of analysis will give opportunity to the Safety Manager to start to think how to stop these scenarios to develop in his system. In addition, he must consider how to eliminate or mitigate the consequences, if any

---

[4] In Russian TRIZ stands for **T**eoria **R**esheniya **I**zobretatelskikh **Z**adatch.

[5] Do not be confused by the usage of the "failure" and "hazard" with the same meaning in this sentence. In the FMEA, failures are treated as hazards which are used to calculate risks through the RPN (Risk Priority Number).

[6] By my humble opinion, the FMEA is the most used method for Risk Assessment and Failure Prevention in the industry in the World. It was first time used by NASA, which later transform it to FMECA (Failure Mode and Effect Criticality Analysis). Today, the FMEA is accepted and used by all automotive industry.

of these scenarios could develop, simply because "The Attackers are always better than the Protectors"!

I was using above the Safety Manager, but things regarding Safety in the Risky Industries are quite different. I have mentioned that we need a Team to produce the List of Hazards and we also need a Team to implement the AFD. Actually, we need two Teams: Blue Team and Red Team.

The Blue Team is a team which, actually, takes care to protect the company from all adverse events in areas of safety. The Blue Team are, actually, the employees who work in every Safety Department in any company in the Risky Industries. Their job is to establish, implement, and maintain the SMS inside the company (Protector).

The Red Team is a group established within the same Safety Department, but they are Attackers: They will try to find a way how the system inside the company will fail. All its ideas, how the system would fail are communicated to the Blue Team, which will try to build defenses against these attacks. So, although, it is the same company and same department, these two teams "fight to win a battle" with the intention to help the company to improve itself in the area of Safety. Both Teams are, very much, effective in helping companies to broaden their problem-solving capabilities in the Safety area.

The philosophy of this situation is described by the old adage: "There are no unconquerable fortresses! There are only bad Attackers!". Whoever is "defending the fortress", the Attacker can always find a way to be more successful. Simply, the Protector's way of thinking cannot be aligned by the way of thinking of the Attacker.

Let's be honest: This could work only in the companies where the Top Management is creative and open-minded with capability to accept the polemics and critics and new or novel ideas.

The problem with the AFD is that it is like dealing with Art: Whatever music song is produced, a novel written, or a painting painted, there will be always infinity number of music songs to be produced, novels to be written, or paintings to be painted. So, it can be an exhausting operation without end.

Anyway, in my humble opinion, this is pretty much a good way to fight not only the BSe but also most of the "safety animals" in the Risky Industries.

9.7 VIDEO GAMES AS TRAINING AND/OR ANALYZING TOOL

Reading different books and articles in the past, I found somewhere the idea of using video games as training tool, and I was excited how good the idea is. Further investigating, I found the introduction of "Advances in Terrorism Risk Analysis", published as a Special issue of *Risk Analysis: An International Journal* in June, 2019. The introduction was written by L. Anthony Cox, Jr. and Michael R. Greenberg. There you can find information that, in the fight against terrorism, Dr. Vicki Bier[7] in her researches, already uses a game theory to understand how attacker prepare for the attacks (to choose a target) by taking care for the defender's actions and weaknesses.

[7] Ms. Vicki Marion Bier (PhD) is an American. She is researcher and decision analyst dealing with risk management, disaster preparedness, anti-terrorism, etc.

So, the game theory associated with the video games is already used to get better knowledge about the adverse events and how to stop them. If all these things found its application in anti-terrorism, why not use them in the Risky Industries to improve the Safety?

There is already something very similar in aviation and it is a use of simulator for pilots and for ATCOs training, but using video games for safety training is really something very much brilliant. Some, maybe, do not appreciate this idea, and I can agree with that, but to learn from the real-life events is too costly. My father's older cousin used to say: The best school is if you study from your mistakes, but the scholarship is extremely high!

It is important to understand that most of people do not learn by studying fancy books with mathematical models, but they do it through experimenting in the real life. The training with video games could contribute very much in this learning area.

Not only for training, the video games with different scenarios could be also used to analyze the human behavior in the particular abnormal situations. In the context of the BSe, it cannot help to predict them, but it is clear that in the same scenarios the humans can behave in some improper way which can trigger the BSe to happen. In the situations of dealing with the BSe during and after they happen and in synergy with the AFD (explained in the previous paragraph), the video games can be excellent tool, not only for training, but also for analysis. Only by "simulating a true situation" through verified scenarios and experiencing them in the realistic way we can prepare better to handle all "safety animals".

So, if the company decides to implement in their safety defense the AFD, then for each possible scenario, a video game can be produced to analyze the responses. In addition, the same scenarios (improved by the results of the analysis) can be used to train the employees for appropriate responses.

The gaming computers can be used for these video games and the virtual reality will be the area which can be appropriately designed for different scenarios. The possibilities are huge, but the real question is: Is there any company which would invest in the design of such video games? And more important: Is there any company in the Risky Industries which would buy such video games to train their employees?

I am very suspicious about that, but the idea is really exciting and, in my humble opinion: It is an excellent idea!

Maybe it can be accepted by the scientists dealing with Safety…?

Not really…

They are mostly dedicated to modelling, but using video games in the area of Human Factors, I do believe can provide valuable information how to improve and how to train the human response.

9.8 REAL-TIME MONITORING

For the simple question "What is real-time monitoring?", I have a very simple answer: Driving a car is a real-time monitoring!

Shocked?

You should not be…

Driving a car is dynamic situation because there are plenty of variables, some of them internal and some of them are external.

The internal variables are the car, the driver, the passengers (if there are any inside), and the interactions between them. Driving a car safely, will depend on the conditions of the car and the conditions of the driver. If something is not "adjusted" in between the car and the driver, there is a risk that something bad can happen. From the point of view of the Safety Management and the Risk Assessment, we can easily calculate the risks for the driver and the car. The number of incidents and accidents which happened in the past will give us the frequency of the hazard of something to be wrong with the car (fault or failure) or with the driver (error or mistake). For each of these hazards, we can calculate also the severities based on the history of the previous driver's incidents and accidents. This will give us the risks.

From the external variables, I would mention the environment, the quantity and the quality of the traffic, the weather, and the road conditions.

The environment is connected with the study of the culture of driving and, later, the maintaining the rules of the traffic. You may say that humans are the same everywhere, but I would disagree: I have driven a car in more than 10 countries and, believe me, everywhere was different. Even the teaching lessons for driving a car differ from country to country. As simple example I can state that in Belgium you do not need to attend a certified driving school to go to the practical exam. In most of other countries, attending a certified driving school is a Regulatory requirement.

The quantity of the traffic depends on the number of the cars on the road when we drive the car. I can mention here also that the quality of traffic is connected with the types of the vehicles which are on the road. Of course, the presence of the trucks will affect the speed and visualization of other cars. From another side, the expensive and fast cars on the road will shorten the driver's time to respond to the possible adverse events.

The weather, associated with the road conditions, is also a very important contributing variable. Simply, the way of safe driving is, pretty much, dependent on the weather conditions. The rain, snow, wet roads, fog, etc. and not adjusting of the speed of the vehicles to the road conditions in such cases, are just few of the problems responsible for increasing the number of incidents and accidents.

So, whenever I drive my car, I monitor the traffic and the environment all around me and, accordingly, I adjust the car speed, the steering wheel and the use of the breaks to the situation.

And that is what needs to be done in every SMS in the Risky Industries!

We can establish any type of management system, but its real value is hidden in the monitoring of the risks. So, in the case of the companies in the Risky Industries, the monitoring of the systems (processes, operation, activities, etc.) is of utmost importance. Whatever the monitoring is, it must be holistic. Do not forget Chapter 1 of this book: There are explained all reasons (Randomness, uncertainty, Chaos, etc.) why we cannot predict the future events. In the situation where the prediction is not possible, the dedicated and holistic real-time monitoring could register the adverse event early, and it will provide more time to deal with it.

In the Risky Industries, there are two areas of real-time monitoring: The first one is monitoring of everyday operations (processes, activities, systems, etc.) and the second one is monitoring of the performance of the SMS. The first one will give us information how our company works and the second one will give us information how the company (humans, assets, environment, etc.) is protected.

Whatever we monitor, it is important for any abnormality to be reported as soon as possible. It will affect the effectivity and the efficiency of the response and the cost of the response and the consequences. If the fire is reported as soon as it is noticed, it is clear that it will need less resources and less efforts to extinguish. Also, if the reaction to the fire is fast and not delayed, the damage from the fire will be smaller.

But it does not stop here…

The real-time monitoring is useless if the corrective actions are not put in place immediately. When an employee who monitors the system notice something abnormal, he must rise an alarm immediately. It means that corrective actions must be implemented by the maintenance employees as soon as alarm is registered. The employees who are in charge of the maintenance must be knowledgeable, skilled, and experienced and must be trained to react to any deviation of the system from its normal behavior. The procedures on what to do, if something go wrong, must be in place and must be executed immediately, if necessary!

OK, there is a "need for speed", but there is always a "but".

There is an adage in aviation which says: If something unexpectedly goes wrong, order a coffee! The point with this adage is that, in highly controlled and "duplicated" industry as aviation, there is no need to be faster than needed. It means, that you need to be fast, but of utmost importance is to understand what is going on!

In the Risky Industry "attacked" by "safety animals", this must also prevail. Your real-time monitoring must warn you to investigate what is going on and if you really understand, there is a "need for speed" in your reaction. If you do not understand, then whatever you do, there is a chance for the action to be wrong.

It is clear that "bad things happen to good people", so we are aware that not always can we stop the adverse events. A good Safety Manager will embed in its SMS the contingency plans to limit the damage and back-up plans how to continue or recover the process (operation, activity, system, etc.) in the case of incidents or accidents. But whatever we are doing, the real-time monitoring is of utmost importance for any "safety animal". It is worth to say that this is very much important especially for the BSe and for the GRe.

The real-time monitoring is affected by the complexity of the systems used in the Risky Industries. Trying to simplify processes will not work there, but there is a possibility to "simplify" the monitoring. I use quotation marks because the word "simplify" in this case does not mean providing a real simplification in regards of the simpler monitoring equipment or cheaper sensors. It means that we will monitor not the process itself, but few critical phases of the process together with the process output. This will help us to notice any irregularities in the process in early phases, so we can react accordingly and in timely manner. This could prevent any "safety animal" to happen and, especially, it could contribute very much to eliminate and mitigate the consequences.

This recommendation, actually, points to another very important aspect of monitoring. By monitoring the system, we are gathering precious knowledge about its behavior and it will increase our confidence what to do in the situations when the adverse events happen. The monitoring, in this case, could be assigned as on-job training for the system in use.

The real-time monitoring is used, also, to assess the implemented SMS in the company. It provides data to calculate compliance of the real events with the KPIs (Key Performance Indicators) in the companies. If the real-time monitoring's data give results which are not in accordance with the established KPIs, something is wrong with our implemented SMS. It must be investigated, and particular preventive and/or corrective measures must be designed and executed. Using the example with the car driving, if the driver notices some irregularities in the car behavior, then he must undertake some decisions and actions which could even go in the direction to stop the car and call a mechanic. If the brakes are not working, stopping the car is an extremely wise decision.

There is another subject which is connected to the monitoring and I would call it "thinking in advance" monitoring.

The Quality is secret, but the Safety is not: It is public!

If companies are reluctant to disclose information regarding the quality of their processes and their products, within Safety, the situation is quite different. There is a Regulatory requirement to report any safety-related events to the State Regulators! Those Regulators will collect all information regarding such events in their countries and, at the end of the year, they will send the collected information to the international Regulatory Bodies. These bodies will collect all this information from different States and they will produce report on the safety events for the previous year in the World. This report will be distributed to the States, which will distribute it to their own companies (subjects in the particular industry).

So, this is another aspect of safety monitoring which is slightly different from the example regarding real-time monitoring of beginning of this paragraph: Driving a car. The point is that monitoring must not be limited to the ongoing processes, but it must include also the adverse events which have happened to some other places, some other companies, in some other States.

In the finance, there is a name "risk landscape" which is used to describe specific conditions of the operation of particular companies. It is clear that there, even in the same industry (banking, insurance, investments, etc.), different companies will maintain different operations. The overall risk, considered in the scope of each individual company's operations, is named as "risk landscape" for this company. In simple words: Different companies – different "risk landscapes".

The reports distributed by international Regulatory Bodies in the Risky Industries may not be considered just as proforma reports. They need to be treated as textbooks and whatever is there (adverse events which we never assumed it could happen to us) needs to be analyzed from the aspect of each individual company (or from the aspect of their "risk landscape"). It must be done by taking into consideration the present reality in the company, its employees, its operations, its processes, and its environment. This is "thinking in advance" monitoring: Some of the bad things did not happen in our company in the past, but we must investigate possibility is there any chance that they will happen in the future?

Yes, I agree: It is a tantalizing job, but only for the beginners or the ignorant guys!

The experienced and knowledgeable Safety Managers will not need so much time to understand what is behind the safety events which happened on some other places and what could be the impact of these events to the situation in its own company. These reports on the safety events are data that have extreme value, not only for calculating the risks (frequency of happening and severity of consequences) but also to prepare your company for similar situations. For the elimination and mitigation of such risks, these reports provide a lot of data, which must be analyzed in advance.

A critical thinking, associated with open mind and particular attention to the details, is fundamentally required for such an operation. It is essential to understand the events which are in such reports and to understand how they can be transformed and would materialize in your reality (your company).

Knowledge gathered from this analysis and measures produced today could literally save lives tomorrow!

9.9 REVERSE STRESS TESTING

There is something used in the industry which is known as Stress Test, but it cannot bc implemented always, especially not for the equipment testing. Stress test is a method of testing when we would like to establish critical point of the system survivability. It means, to find this point, we need to destroy the system. And this is very much expensive. So, that is the reason that in industry, the stress test is used mostly for software.[8] There, whatever damage is caused during testing, the back-up of the software and the data always can restore the system in its previous configuration.

In the title of this paragraph, it is mentioned Reverse Stress Testing (RST). It is something which is very much used in the financial institutions (banking, investment companies, etc.). They have many methods to calculate the business risk and the Reverse Stress Testing is very much popular there.

The Reverse Stress Testing is a testing of the overall company, where their established business model (or some new change in operation or new investment) is assessed for particular bad scenarios, possible circumstances and conditions that, if show up, would endanger the model and make the company unsustainable. This is a testing which shows up the vulnerabilities of the companies, mostly in the financial aspect of their business, caused by the unpredicted events in their future operations.

In the Risky Industries, the RST could be associated by the testing of already established SMS. This is covered by the Regulatory requirements for the SMS, where there is need to examine the SMS performance. I would not go into detail how it could be done. Simply, each company in the Risky Industry will decide how they will examine their safety performance.

In addition, it can be used by the companies in the Risky Industries to calculate survivability of the companies if some accident happens to them. These companies always complain that maintaining the SMS is expensive job and the RST could really provide real picture how much the accident costs. I am dying to see what is the ratio between the costs of maintaining a proper SMS and the costs if accident happens…

Whatever is explained in Sections 9.6 (Anticipatory Failure Determination (ADF)) and 9.7 (Video Games as Training and/or Analyzing Tool) can be used for the RST.

[8] The Crash Test in the automotive industry is a Stress Test for the cars, and it is regulatory requirement. In the test few cars of particular model are tested and destroyed during the test.

9.10 CHANGE MANAGEMENT

The real-time monitoring is very much important when, in the company, some change needs to be introduced. The change can be anything which could affect the processes (new equipment, new technology, new procedures, new employees, etc.). This is something which is requested by the regulation in the Risky Industries: There is a requirement for Change Management which shall be implemented before the change takes place.

Whatever the change is, it will usually solve some problems (eliminate or mitigate some of the risks), but it will also introduce new hazards and risks and, eventually, it will change the existing ones. So, the Hazard Identification must be executed again and all determined risks should be evaluated again, together with the newly identified ones.

The very much important aspect of the change is, also, the explanation to the employees why the change will take place. The humans are very much reluctant to the changes, so each change must be considered as a Human Factor. Proper explanation based on arguments and facts will decrease the reluctance and even may encourage the employees to be proactive in the change implementation.

So, the Risk Assessment of all aspects of the change should be done in advance. If there are some new or novel hazards, they must be the subject of preventive or corrective measures. In addition, emergency procedures and back-up plans and contingency plans should be prepared. The wise Safety Manager will always insist on the testing of the change before the implementation. This must happen because, you must understand that each change is a GRe and if it is not introduced properly, it could result in catastrophic consequences.

My humble experience tells me that, in reality, the companies are very much bureaucratic regarding the changes and mostly, they follow administrative rules by neglecting attention in advance to the change.

If the company is not fully prepared in advance for installation of new piece of equipment or introduction of new technology, then the happenings of any "safety animal" is highly possible. With those things, there are always no enough data to predict every aspect of the change.

There is another one very important aspect of the Change Management in the cases when change is introduced due to new trends on the market or new regulation. In such a situation, the program for introducing the change should be realistic, done in advance, and the progress must be carefully and thoroughly monitored.

So, be careful when you implement Change Management! There is a possibility to some of the "safety animals" to show up from, literally, nowhere.

9.11 PREDICTIVE MAINTENANCE

In Section 2.8 (The Equipment and the BSe in the Risky Industries), I have explained that there is a method to calculate probability of the faults for the Equipment in regards to the BSe, but there is something else which could help fighting the "safety animals" caused by their equipment faults: Predictive Maintenance.

For the time being, most of the companies are dealing with Preventive and Corrective Maintenance.

Preventive maintenance deals with changing the parts in the Equipment which are subject to wear. After particular time, the filters and some parts of my car shall be changed in the workshop. If not changed, they could cause other faults which could endanger lives and damage assets (my car, for example).

It is not necessary that the faults caused by the wear parts need to be so dangerous. It is enough if the faults cost me more money to fix them than the preventive changing of the parts in the periodic intervals. In this process, there is no need to have particularly knowledgeable or skilled employees. This is not so creative job at all, because every step in the process is well defined by the procedures. But it is very important for the employees to be trained in the executing of the procedures.

The corrective maintenance deals with fixing the faults which happen on the Equipment due to different reasons. The corrective maintenance usually needs more knowledge, skills, experience and different attitude from the employees who are responsible for this job. The people dealing with corrective maintenance in the industry are like medical doctors for the Equipment. Having in mind that the Equipment does not speak to them, they shall be knowledgeable about the processes, the systems and their mutual interactions. They usually use different types of instrumentation to understand what is the cause for the fault. The complexity of today's equipment is time-limiting factor to do their job as soon as possible.

The third type of maintenance is Predictive Maintenance. It is actually upgrade to preventive maintenance.

The predictive maintenance should provide prediction when the parts need to be changed before the faults happen. There is a term "Remaining Useful Life" (RUL) which will help to decide when to change the critical parts. As such, it is connected to reliability.

The predictive maintenance is a type of maintenance which uses additional monitoring (by humans and/or sensors), data gathering, data processing, and calculations to provide knowledge in advance about wearing of particular parts in the Equipment (Figure 9.1). That is a method to provide a benefit of extending the time between changing the parts due to wearing. As such, it is based on science and engineering regarding the fault preventions, but it mostly contributes as an economic category. Actually, it helps to save money and resources by extending to the maximum the time interval between changing parts in the preventive maintenance. Extending this time will extend the time of production of products, which means: "More time – More products – More profit". This is considerable gain for the companies which use a serial production of products.

The Monitoring shown in Figure 9.1 is different than real-time monitoring already explained in the previous paragraph. There, the monitoring is implemented in the real time by humans, monitoring indicators, or sensors of processes and, as such, it can provide immediate reaction, if there is any abnormal event. This type of monitoring

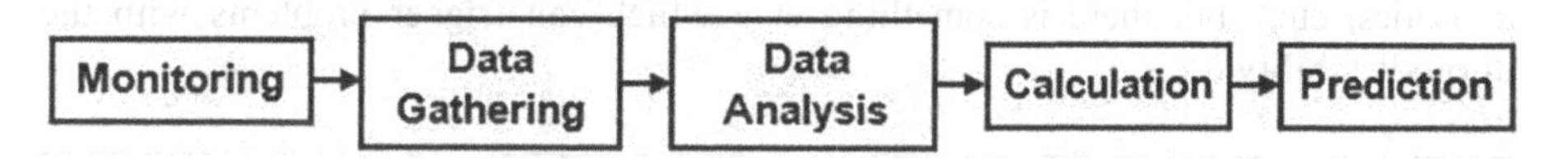

FIGURE 9.1 Steps in predictive maintenance.

applies to the operations of the system which provide production of the products and, if there is a fault, this type of monitoring will raise an alarm. As such, we can define this type as "event monitoring".

The monitoring in the predictive maintenance (Figure 9.1) means: Following, in regular intervals, the shape or the health of particular parts which are subject of preventive maintenance and, as such, if not changed on time, they can fail and cause accident or incident. It can be also defined as a "condition monitoring", because we gather data regarding the condition of the parts.

The data gathering can be done by Humans or automatically (by sensors) and the data gathered can be manually or automatically transferred to the system for data analysis. The data gathered can be different types of data such as vibrations, strange sounds, oil cleanness, high or low temperature or pressure, high humidity, etc. The change of the values of these data could (statistically) help to predict a life span of the particular parts of the Equipment.

It needs to be stated here that these data are not necessary analyzed in the real time, simply because the data which will provide better information must be historical data. It means the condition of the equipment must be followed for all time, from putting the Equipment in operation, and the trends in the periods of maintenance must be established. Of course, having in mind that there is need of huge amount of data, there is also need for advanced processing techniques for this "big data". The data analysis should provide diagnostics about conditions of the parts.

The results from the diagnostics, based on all these changes of part's conditions, could provide data for calculation of the time, which can serve as prediction of possible faults of the Equipment. This is actually prediction[9] regarding a Remaining Useful Life (RUL) that will provide information regarding a time for the next preventive maintenance. Of course, time to change the parts will depend on calculated RUL and, as such, it must be always shorter than RUL.

The processes of the diagnostics (from data analysis) and prediction (from calculation) are interrelated because the prediction is tested in the reality, and this testing provides more data for the diagnostics. The newest directions of the research and development in this area move towards the Artificial Intelligence (AI) and Machine Learning Decision-Making Process[10] (MLDMP).

The predictive maintenance deals only with the Equipment and as such it is still dependent on the human decision-making process. In addition, the data analysis is based on statistics with all possibility of wrong data analysis and other inconsistencies of the statistical analysis.

9.12 MANAGEMENT SYSTEMS

It was mentioned in Section 1.5 (Reasons for Unpredictability) that the unpredictability can arise also from the non-linearity of the systems (processes, operations, activities, etc.), but there is something else which can trigger problems with the unpredictability…

[9] In some literature you can find a word "prognostics" for this process.

[10] MLDMP is already used in some hospitals for diagnostics of human diseases.

It is clear that each management system is built by congregation of Equipment, Humans and Procedures. The issue here is that the relationship between all these three building elements is mostly non-linear and, as such, they need to be monitored and maintained to provide particular control over processes of any management system. Of course, due to the possible consequences of the adverse events in the Risky Industries, the use of the Safety Management System (SMS) there, is very much important. The defining the non-linearities of the implemented SMS is not easy job and no one is doing that, so there is a need to monitoring the SMS.

As it can be noticed, it is imperative to monitor all your processes (operations, activities, etc.), but in addition you need to monitor also your management systems. Having in mind that management systems manage the Humans, it is clear that some human errors/mistakes can arise from the incapability to understand the non-linearity of relationships between the Equipment, Humans and Procedures, in addition to everything else. This is something which I call "forgotten aspect of the management in the company".

The main point with any management system is that it is a live system!

Whatever monitoring we do on the effectiveness and efficiency of any management system, in the case when something is going wrong, the particular action must be triggered. The action can be preventive or corrective, same as the measures which we implement as part of the Risk Assessment. It means that the system must bc ready to adapt to any of the changes of the situation or environment, inside and outside the company.

9.13 CONTINGENCY PLANS

One of the parts of dcfinition of the BSe is that they are unpredictable. Having this in mind, trying to predict them, it is not a wise action. Instead of that, it is pretty much wise to prepare in advance to deal with the consequences. Nevertheless, very often, even the consequences cannot be predicted. This is something which is very much used as defense from the earthquakes. We cannot predict the consequences, but we can provide resilient building and assets which will eliminate or mitigate the damage from the consequences.

How the Safety functions today?

By thinking in advance!

We prepare the List of Hazards where all possible hazards are identified in particular area (company, organization, domain, etc.) in the Risky Industries. For each of these hazards, we determine the likelihood and severity of consequences transforming each hazard in particular risk. We try to eliminate each risk, but it is not always possible to do it. For those risks which we cannot eliminate, we try to mitigate them (decrease the likelihood or decrease the severity of consequences). Of course, even these mitigations will not always work, so there is need to provide contingency plans. These are plans which will provide the answer on the question: What will we do if the bad things happen?

There is another reason to have contingency plans prepared in advance…

Imagine that you monitor your system (operation, process, activity, etc.) and you notice that something goes unexpected with the system behavior. You do not know what it means and you try to understand what goes on. If it is a BSe, it never happened before, so what are the chances to understand what happens? If it is a GRe, will you understand what kind of harbinger this is? You will try focus on the event, but the IG could prevent you to miss the main point... So, in all such situation, the emergency procedures, built into the contingency plans, could help to react until you do not understand what is going on.

The contingency plans are plans prepared in advance by the management and they actually define the actions which need to be executed, if the operation (process, activity, system, etc.) deviate from its normal behavior. They are common in the Governments and businesses to prepare for fast recovery after disaster happen or if there is some other kind of economic disruption. When used by the Governments, there is usually a regulatory requirement to produce such plans. The contingency plans are used also in the project management community, mostly as plans how to continue, if the things do not go as planned in advance. Most IT companies are taking care for their databases by back-up and by contingency plans. The loss of data could be a disaster for any such a company.

In the Risky Industries, there are also Regulatory requirements for the companies to produce such plans. The contingency plans in the safety area shall be structured in such a way to provide the appropriate response by trying to eliminate, mitigate, and/or confine the consequences. The confinement is very much important because many of the adverse events can " build up", if not confined. In addition, the contingency plans shall define the possible means for continuation of the operation (to provide business continuity).

The contingency plans must contain the rules about the chain of command and procedures of what to do, with clear description of accountabilities and responsibilities to every employee.

It is understandable that the contingency plans are not simple because there are many risks and there is need for particular resources to execute them in a timely manner. Anyway, it is important to keep them simple. When anything bad happen, the people are confused (chances that they will panic are big) and the complexity of the plans could be a big obstacle for proper immediate response. In such cases, the simple steps for the activities and for the flow of information are essential to provide good execution of the plan.

When the contingency plans are prepared, there is need to share the drafts with the employees. This is very important step because the comments of the employees regarding the contingency plans could provide valuable information about sustainability, effectiveness, and efficiency of the plans. After reconsideration of all aspects of the comments and their importance regarding implementation of the plans, it is necessary to provide particular training for employees. This training should be managed to provide on-site (reality) test for the plans. Of course, this is not always possible or feasible, but it is important to test the plan. Having appropriate check list (for tasks and resources available) for testing a plan is a wise step.

Having in mind that the contingency plans will not be used very often, there is big chance to be forgotten, so particular refreshment (theoretical and practical) is wise to be implemented at least once per year.

Having in mind that the working environment in the Risky Industries is dynamic, the Safety Manager must take care for periodic review of the contingency plans in his department. Do not forget that contingency plan is a "live" document. The periodicity (monthly, annual, etc.) for the review can be established after consideration of the changes in the operation, complexity, environment, structure or whatever else happen as new, in the company.

Whoever is producing the contingency plans, he must be open-minded. There is no simple procedure how to do it, but having good understanding of the structure and risks in the company, trends in the industry, technology used, education, culture diversity, training and experience of employees, are essential precondition to produce good contingency plans.

9.14 REGULATORY BODIES

Regulatory Bodies are an important reality in the Risky Industries. Due to profound and wide-spread effect which the incidents and accidents could produce, there is a need for international regulation based on cooperation and coordination, so all these Regulatory Bodies are usually international organizations. Sometimes, these organizations are agencies of Organization of the United Nations and sometimes these are just bodies established by multilateral agreements between the States.

The States, which are members of these organizations, are required to establish their own national authority (State Regulatory Body: A State Regulator). It can be a ministry, an agency, an administration, an institute or whatever the State reckons is appropriate to take care for the international and State regulation in the particular Risky Industry. The first responsibility of the State Regulator is to transpose the international regulation into a State legislation. In addition, they may produce their own regulation, which must not be in conflict with the international regulation.

The level of connection of all these three subjects is shown in Figure 9.2.

As it can be seen from Figure 9.2, all three subjects communicate between themselves, but companies do not communicate to the international organization. For them, the State Regulatory Body is the authority which will require the compliance with the regulations.

The international organizations oversee the State Regulatory Bodies by the use of different tools and methods, but mostly, it is done on the basis of the evaluation of the regular reports about safety-related events which they receive from the State Regulatory Bodies. These reports are not the only form of overseeing the States and there are few other forms (approving regulations and documentations, assessing training/education, audits/inspections of the State Regulatory Bodies and the companies, etc.). One of the very powerful tools is a regular (periodic) audit. These

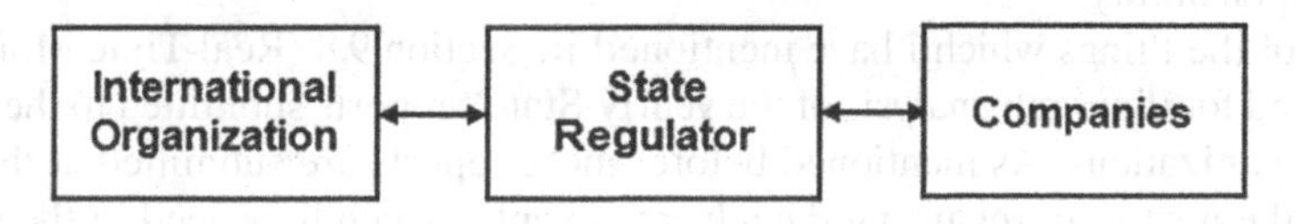

FIGURE 9.2 Relationship between Regulatory subjects and companies in the Risky Industries.

audits are done to the State Regulatory Bodies by a team of auditors which usually checks the implementation of the regulations and, most importantly, they check, is everything which is in the regulation, implemented in the reality. The team of auditors could be established from the employees inside the international organizations or it could be established from the pool of external experts, already trained and approved by the organization. These pools of the experts are cheaper solution and it prevails in most of the international organizations. The experts do not need to be employed in the organization, but they are paid for their engagement during the audit.

In the State, the State Regulatory Bodies in each area have responsibility to oversee the companies and to assure, first, themselves, and later, the international organization that everything is in compliance with the regulations. These State bodies operate in the area between the international organizations and companies and they must deal with both of them. Their job is to oversee the implementation of the regulation inside the companies and for such a purpose they use also different tools and methods. Most powerful is the audit, which, depending on the situation and industry, is done at least once per year.

In the scope of each individual State, the State Regulatory Bodies are accountable for the situation of particular Risky Industry in the State and all companies are responsible their operations to be safe. As such, the companies are at the "end of the chain". The companies strive to make a profit without endangering the humans, assets, and environment and by assuring the State Regulatory Bodies that the compliance with the regulation is achieved.

It is very much wrong to assume that the companies must abide to the regulation due to regulation itself! It is wise if they assure first themselves that their processes (operations, systems, activities, etc.) are safe, after that they shall try to assure the State Regulatory Body.

Unfortunately, this is not going always as planned and as required. The reasons are the "safety animals" which affect the companies and their management. I already have spoken about all these issues in this book, but I would like to point here that these "safety animals" are affecting mostly humans and humans are part of the Regulatory Bodies also. It means that the Regulatory Bodies are very much prone to all these failures with one very important notion: Having in mind that the State Regulatory Bodies take care for overall industry in the State, the "safety animal' which can pop up inside the State Regulatory Body could affect all the industries in the State.

Having in mind that the employees in the State Regulatory Bodies are not well paid as in the private companies, not always are the most competent guys employed there. And having in mind that they are in charge of the producing the atmosphere and the regulation which affect all industries in the State, I hope you can understand their responsibility.

One of the things which I have mentioned in Section 9.8 (Real-Time Monitoring) is the need for thorough analysis of the yearly State's reports submitted to the international organizations. As mentioned before, these reports are submitted at the beginning of the next year, regarding the adverse events, which happened in the previous year. The primary duty of each State Regulator is to analyze the data gathered from the companies in the state. This analysis shall show how the things are going in the

State regarding the safety in this particular industry. If calculated KPIs are better than the State-established KPIs, than things are going well. But if there is data which shows impairment of the situation, the duty of the State Regulator is to investigate the root cause and to propose measures to improve the situation.

In addition, the yearly report received by the international organization must be thoroughly analyzed by the State Regulator and if there is something which will request the change of the State regulation, it must be done in cooperation with the international organization. The companies, sometimes, cannot solve the adverse events which are reported in these yearly reports, so it is normal if they would need help from the State Regulator.

The most important thing which must be understood by the State Regulator and the companies is that this is not a Master–Slave relationship. This is a relationship of cooperation and mutual understanding where both subjects, in synergy, try to provide safety of their people, assets, and environment. They are just "two pages of the same sheet of paper" and, as such, they cannot exist independent from each other. Any misunderstanding of this aspect of the regulation is good terrain for dealing with "safety animals" of any type.

In the scope with the Master–Slave imposed relation from many of the State Regulators, I would like to warn them that they are not excused from any investigation or abolished from any guilt. Let me remind you on the example stated in Section 6.4.2 (Defense Lines of the Regulator) in this book. There, I have presented one single example from aviation: In USA, the State Regulator is Federal Aviation Administration (FAA), but when there is an accident, the body in charge is National Safety Transportation Board (NTSB). It means that FAA is also under investigation by this body.

9.15 SURVIVAL PSYCHOLOGY

The first human reaction of the BSe and GRe is shock, which most often results in panic. The panic is result of the fear not knowing what just happens, what to do, and how to proceed. Both, the fear and the panic will not help with the fast response, even in the case when there are already measures to handle the adverse events and to prepare terrain to deal with consequences.

A lot of psychological investigations were done, but most of them dealt towards understanding and providing medical therapy to the survivors after the disasters happened. The effect of all these investigations was medical and legal recognition of "Post-Traumatic Stress Disorder" as registered illness.

Unfortunately, it was a "reactive" approach, dealing with the consequences. But what is really necessary is to be focused on understanding and appreciating the psychological functioning of the humans when actually the sudden disaster happens "from nowhere". Addressing the human functioning under huge stress could help to understand what kind of immediate measures can be implemented to eliminate or to mitigate the adverse events and their consequences afterwards. Starting early with these measures will save lives and money, especially in the area of the Risky Industries.

The new initiative in this area is known as Survival's Psychology.

9.15.1 Survival Psychology for Individuals

Survival psychology is a discipline, in the scope of the psychological science, which deals with the research about human behavior in life-threatening situations. So, it is not what to do after the life-threatening situation has passed but what to do before and in the moment when the life-threatening situations happen. The area of investigation is mostly based on psychological and hostile environmental conditions which are capable of destroying human beings. This psychology established a rule that the humans who survive the adverse events are not survivors until their fully functionality is not established again. The survival psychology assumes that the post-survival situation with the humans can be extremely improved if particular training and measures are provided to the humans before the bad things happen.

The level of the fear and panic in the humans is highly individual. Not all of them will feel the same fear and show same panic in the same situations. The fear is primeval and the main point with this highly individual reaction is that it will produce panic which will make you to be paralyzed or to run away. As such, it will not work for the Risky Industries in the same way as in other industries because there is more than a simple survival. The paralysis or the running away from the scene of the adverse event will make worthless all efforts put in place for preparing the emergency procedures, back-ups, and contingency planning. The reason for such a reaction is that adrenaline which is secreted in such a situation from the adrenal gland will affect the part of our brain which is dedicated to complex thinking. The human capability of complex thinking will not be available at that moment, the humans will be incapable for requested response and that is the reason that the individuals must react unconsciously, or in other words: Instinctively! Instinctive appropriate reaction can be provided by training in advance.

The training is based on the repeating the survival operations based on procedures until it does not become an automatic reaction. This is something which is implemented into the military training. The periodic regular training is something which can build up this automatism: Every soldier is trained to react automatically in the life-threatening situations in the battles.

Another important point is that the fear and the panic can be contagious! If one of the staff started to panic, the panic could be very easily spread around. In such a moment, there is a need to have a person who is present there and who will not panic. He can overturn the panic by his reasonable attitude. In general, the panic will not be spread if someone undertakes the situation and, authoritatively, starts to command activities which make sense. In this situation, the staff which already accepted panic of one man will realize that things are not so bad, simply because, there is someone who thinks that there is no need for panic. They will shift their attention from the panic guy to the reasonable guy and there is hope that the adverse event, and the damage caused by it, can be controlled.

The individual investigations regarding the survivability of the individuals showed few factors which can improve survival. Let's go through five of them:

1. **Knowledge** – Having knowledge how the operation (process, system, etc.) is conducted will be highly beneficial with the adverse event which just happened. It could suppress the fear because it can help to understand what

has gone wrong. It means that the adverse event can be analyzed very fast and the solution what to do in these first moments can be found. Whatever decision needs to be made, the knowledge about the system will help to find a good one. In addition, you will need skills necessary to deal with the damaged process or equipment and have in mind that the skills are built with training.

2. **Human Condition** – The condition of the humans when the adverse event happens is an important factor. Being tired (sleepy, hungry, thirsty, etc.) will worsen the reaction. That is the reason that some employees in the Risky Industries are obliged to take care of their mental and physical situation: They must be in excellent condition when they are in the company! When I was child, I remember, there was crash in two trains in Yugoslavia and this is the worst disaster which happened ever in this country before it stopped to exist. On 30th of August 1974, the international train from Belgrade to Frankfurt entered Zagreb's railway station with speed higher than 90 km/h. Due to the speed, during the crash, the train disassembled itself and 150 passengers died. More than 100 from the train and at the station were injured. Later, investigation showed that the driver and his assistant worked approximately 300 hours that month. In addition, before boarding this train that night, they worked 51 hours without a break before. The train was late and the driver wanted to compensate the time by entering Zagreb Railway Station at a higher speed. Obviously, he overestimated his capabilities to maintain a safe entering.

   There are plenty such adverse events caused by "wrong" human condition. Healthy body, together with a good mood, can handle better the stress and the fear in these situations. As an example from aviation, I can mention that pilots and Air Traffic Controllers (ATCOs) have additional extra days to take rest;

3. **Environmental Conditions** – Having pleasant environmental conditions will help very much the human reactions. In adverse weather situations (snow, rain, fog, strong winds, etc.), the human reaction will be impaired and results will not be as expected;

4. **Luck** – Nothing without Luck! Someone will ask how the luck can be brought into life-threatening situation and answer is: It cannot be brought! It just happens! There is a video shared on Internet when the car which races on one of the rally races in winter lost control and missed one of the spectators by a few inches. This is pure luck!

   The luck happens if your try to change the things. So, there is no chance for luck if you are just sitting and waiting for your fate. I remember the expression which I have read in one of the Marvel Comics (when I was child) and it says that "luck is helping only to the brave man". The humans must be proactive and this proactive attitude could bring them luck;

5. **Will to Survive** – This is attitude of utmost importance! Without it, nothing from the previous four factors will help! If you are fighter and you are proactive in your life, your chances for survival are better. This is an ability not to be disappointed and be ready to continue, whatever and whenever happens.

> Accepting the nasty reality is important because it will enable you to act by directing your focus on looking for answers about what to do to deal with the adverse event. It is not easy to accept or to deal with this, but it is 50% from the job which needs to be done to survive.

The training is very much important thing in the survival. It creates a memory in the human body, so the muscles react automatically by the pattern of survival behavior which will be embedded into them by the training. The training can help, but not always, so it is wise if immediate reaction of the adverse events in the Risky Industries is connected with the safety equipment. It means that the first human reactions need to be based on the use of equipment.

The equipment is excellent because it does not have feelings and is not prone to human emotions and reactions, so it cannot make wrong decisions. Having automation embedded into your production systems can help very much. It means that instruction "Press the Button!" in the cases when automatic response is part of the equipment could provide a "safe haven" in most cases. If you do not understand what am I speaking about, please remind yourself on those big red buttons with the yellow base which can be found on any piece of equipment (system, machine, power supply, bench, etc.) which immediately disconnects the power and stops the operation.

This is a very rudimentary solution, and more complex systems are using more complex automation to stop the process or operation, but the point is that, although in the Risky Industries there are plenty such systems, the bad things still happen.

And, at the end of this paragraph, we are going to the Darwin's theory of evolution: "Those who are most adaptable - Those will survive". This is applicable to all nature and it is applicable to the humans also. The capability to adapt yourself on the new situations (even it is life-threatening situation) could make the difference between death and life.

9.15.2 Survival Psychology for Companies

The companies will go into bankruptcy if they go out of money. They will go out of money if they spent more money than they earn, but this could rarely happen. The biggest problem is when their business is not "selling". If the product or service is good, it will be sold, but what about situation when the customers start to question the "goodness" of the product or service? Will someone go into restaurant where there was a food poisoning episode one week ago? Will a passenger choose an airline which had report that crash of its aircraft last year as a company failure? Will a woman buy a cream from the cosmetic company which has a track record of using dangerous materials for their products?

These are questions of "one million dollars" and I would not respond to them here…

All that was mentioned in the previous paragraph is mostly dedicated to the individuals. Now, let's think how we can try to produce something similar: Could the managers produce atmosphere in the companies which will support the "survival" of the processes and operations inside the companies when the bad things happen?

I think it is another "question of 1 million dollars" and I will respond to it here: The answer is YES!

So, having in mind the individual reaction, it is plausible to realize that it can be transferred into a group reaction.

If you go on the Internet, you can find many articles which deals with survivability of the companies in the financial area. The specifics there is that most of the problems are caused by money and can be solved by money.

In area of Risky Industries, "the context of the things" is different…

Let me remind you on the White House report on the benefits of implementation of SMS in airlines that was mentioned in Section 6.4.1 (Defense Lines of the Company). There was clearly stated that "thinking in advance" by putting some efforts, which are more based on procedures than on money, could really make a difference. In other words: Preparing in advance will cost you considerably less than dealing with consequences later.

A simple example is the Deepwater Horizon accident which happened in April, 2010. The managers there were worried about daily costs of 200,000 USD and tried to speed up the closing of the drill as soon as possible. These efforts made them incautious about the procedures for closing a rig and, later, when measurement results which were inconclusive, they choose the worst option: Not to find the reason for inconclusiveness! Eventually, many years later, the overall cost paid by the British Petroleum was estimated at 65 billion USD.

In the industry, the ownership of the companies is a big factor of their survivability. As it has been shown, the most of the private companies, especially those with family ownership,[11] have better chances of survive than companies with other types of ownership. It is assumed that the families take more care about their companies and, in addition, they are not keen to undertake risks as other companies would do. In the Risky Industries, the number of the companies with family ownership is not so big which means that most of them are not necessarily good in surviving.

The big companies controlled by a big Board of Directors are prone to survival better than companies controlled by small Boards of Directors. The reason for that is that more people could provide wider range of views on the fate of the company and, as such, it can give more information which will help building better picture of the situation. With more information, the decision-making process is better. Also, with more people on the Boards, the undertaking unreasonable risks will not be possible. It is understandable because with more people in charge, the accountability will be better. Anyway, in general, bigger Boards do not imply better performance of the company.

Those two things above apply to any industry, but the survivability of the companies in the Risky Industries is affected not only by the dynamic business environment or by their ownership and type of governance. It is also affected by the adverse events which can happen to them due to the nature of their industry. That is the reason that all the companies in the Risky Industries are strongly regulated and overseen by the State Regulatory Bodies.

[11] You can find on internet ERC Research Paper No. 1 titled as "Family Business Survival and the Role of Boards" (N. Wilson, M. Wright and L. Scholes) issued at May 2013 by Enterprise Research (www.enterpriseresearch.ac.uk). (last time opened on 12th of March 2021)

The companies cannot think by themselves. They have managers (Top Management and CEOs, General Directors, etc.) who are taking care of the companies. I will speak about them in Chapter 10 (Top Management and "Safety Animals") and here, I would focus on the things which those people should focus to provide survivability of their companies.

The Top Management must strive to provide an atmosphere of professionalism in the company. It means that the humans who are employed there must have four important characteristics. They must have, in particular:

1. **Knowledge** – This is something which can gave them insight into all systems established in the companies. They will understand the inputs and outputs of operations together with the way of transformation of inputs into outputs during production processes. This knowledge can help them to make a difference between normal operations and abnormal situations. This is something which happens in the medicine, but the excellent example is actually a veterinarian doctor. He is educated (knowledgeable) of all normal functioning of the different animal bodies, so he can recognize what is abnormal situation with them.

   Why this is an excellent example?

   Because humans can speak to the human doctors and describe their problems. The animals cannot do that. This similarity of veterinarian doctor with the engineers in the industry is striking: The machines and equipment also cannot speak to the engineers when they have a problem. The engineers must use gauges, sensors, other instrumentation, and equipment (built in the machines or external) to understand what is going on. And without considerable knowledge, this is impossible;
2. **Skills** – The skills are coming after the knowledge, simply because when the knowledge will provide information what is wrong, skills are used to fix the problem. Skill can be a soldering for the electrical engineer (to change elements in the circuits), use of machines for the mechanical engineer (to produce a part from the scratch), etc.;
3. **Experience** – Knowledge makes you good, but experience make you perfect! Knowledge gathered from the universities and the schools is limited to the ordinary cases which can happen in everyday lives in the industry. The interaction between the building elements in the system and the environment are something which can be gathered by experience and, as such, it is precious. Sometimes, uncommon things which happens in the industrial processes can provide insight into something considerably important and it must not be undervalued. For the knowledge, you need intelligence, but for the experience, you need time; and
4. **Attitude** – All three previous characteristics are nothing if the particular human attitude does not exist within a person. The appropriate attitude is a building atmosphere for the professionalism in the companies. This is something which is mostly neglected by the Human Resources Departments. Most of them during interviews completely neglect the attitude of candidates. The employees in the Human Resources are mostly focused on

persons with weak personalities because they can be easily commanded. But can these persons fight for something which they have noticed is wrong?

I have very strong doubt about it! These persons are useless in solving the problems and you do not need them in the Risky Industries. There is simple example about incapability of the Human Resources Departments to provide the appropriate persons for the companies and I will use it here: Steve Jobs. For those who have read his biography from Walter Isaacson, it is clear that he was very a hard person to deal with. So, the real question is: Would Steve Jobs pass the recruitment process in any company? Having in mind that I have changed many jobs and have attended plenty of interviews on both sides (interviewed and interviewing), I am sure he will not pass ordinary recruitment process. And who was Steve Jobs? The man who built the most valuable company in the history of the world: Apple!

The good atmosphere in the companies can be built by the Top Management who will assure employees that their knowledge, skills, experience, and attitude are welcomed and highly appreciated in the company

But this is not enough…

These employees must work as team and each team needs a "conductor" – a guy who will synchronize and coordinate the management team. He should be the "brain" inside the company's "body". This is a person:

a. Who must have feeling about the situation in the company and in the industry, and this is something which cannot be learned at school;
b. Who will know the capabilities of each member of the team and he will know how to use them for the benefit of the company and the employees;
c. Who will understand that his own example will be treated by his coworkers as a standard of behavior in the company;
d. Who will create an atmosphere of progress and common understanding of different views regarding the different cultures and personalities in the company;
e. Who will release the pressure (stress) from the employees in any situation when this pressure is built up;
f. Who will have respect for the regulation and will be familiar with the methods and measures how to provide compliance with the regulation;
g. Who will undertake the burden on his shoulders in the hardest times and will direct the company out of problems;
h. Who will be the Leader (when it is needed) and the Manager (most of the time);
i. Etc.

These qualifications can be achieved by a talent and a lot of self-control and this is something which cannot be found in the books. That is the reason that the Human Resources employees will fail to recognize Steve Jobs as a person who has established the most valuable company in the world.

As we can see later, in Chapter 10 (Top Management and "Safety Animals"), Todd Conklin said that incidents/accidents are caused by the latent conditions in the companies that lie dormant and hidden inside the companies. The Top Management is responsible for these hidden "opportunities" of latent conditions to materialize. So, having a good Top Management is of utmost importance for survivability of the companies (whatever the word "good" means!).

Another aspect of the survivability of the companies is an independent audit. Having independent and impartial audit is of utmost importance for the companies in the Risky Industries. It is important because the insight from outside of the company can be extremely valuable information for the good managers. The auditor from outside has no "memory" of the present and previous situation in the company, so he is extremely convenient to provide clear picture of the company seen as outsider (without all internal errors, mistakes, fallacies, and delusions). The auditor from outside could provide information about non-registered (usually systematic) problems which can be easy solved by the company, if there is proper information about them. As such, he can help to diagnose the problems and the management of the company could propose the measures to fix the problems.

In the Risky Industries, there are scheduled audits from the State Regulatory Bodies at least once per year. There is also a regulatory requirement for internal audits once or twice per year. Anyway, in my humble experience, many of these internal audits are done proforma; mostly they are biased and they are done by the employees who are just formally trained for audit. As such, I do not believe that they are capable to provide a realistic picture about the situation inside the company. If you like the audit to be useful, it must be done by independent auditor or independent body.

And, at the end of this paragraph, let's go again to the Darwin's theory of evolution mentioned for humans in the previous paragraph and paraphrased here: The companies in the Risky Industries which are most adaptable to the dynamic nature of their operations (processes, activities, market, etc.) and in the industry itself are those that will survive. The capability to adapt the company to the new situations (even it is any "safety animal") could make a difference between death and life. The dynamic, complex and, sometimes, unpredictable environment in the Risky Industries demand this capability for adaptation. The humans change their habits, values, and tastes; the markets change the economy; the Regulatory Bodies change the regulation; the States change the laws; etc., and if the companies do not adapt to all these changes, they are in a situation to deal with any of these "safety animals".

9.16 RESILIENCE ENGINEERING AND RESILIENCE MANAGEMENT

One of the things which can help with consequences is the resilience of the system.

In general, whatever is the system (engineering or management) which we use for our operation or processes, it must be adaptable. This adaptation is needed because the system has to deal with possible unintentional or intentional changes of the internal or external parameters which can affect the system. If the system is not adaptable, then it will experience failure. If the system is adaptable, it will continue with the operation and we can call this system a Resilient System.

In the area of the Equipment, there is already a discipline called Resilience Engineering (RE). This is something which has been innovated by the "scientists" who did not follow the activities in the area of Quality in the past.

The father of RE is Dr. Genichi Taguchi, but the paradox is that in many books and articles regarding RE his name is not even mentioned. In my humble opinion, his work was done at the second half of the last century and most of his published books are from the end of that century and the beginning of this century. The RE is something which comes under this name in the last 15 years. Dr. Taguchi was actually pioneer of the RE in the last century, but the name of the method which he pioneered was Robust Engineering. He was the first one who tried to materialize the idea of systems which will be "tough" and "elastic" in the same time, so they cannot be affected with any adverse events caused by any internal or external factors.

It is interesting to mention that partisans of the RE have announced that Resilience Engineering and Robust Engineering are different, but in my humble opinion, it is just apologetics. When the "clever" guys who innovated RE realized that, long time ago before their innovation, there was something called Robust Engineering, it was easy to declare differences.

The next step in the development of the resilient systems could be mentioned an Autonomous system.[12] This is system which has autonomy to change itself with intention to adapt itself on the unexpected changes of the working or environmental parameters. Of course, the autonomous system does not need human intervention to adapt itself, but the level of computerization and automation in the system is so huge, so it can happen easy without human presence.

So, in areas of Equipment, the RE can improve the things if we design our Equipment to be "tough" and "elastic", but with the humans, the things are not so simple. Speaking about the SMS, we can buy the "toughness" and "elasticity" of the equipment with money, but this will not work with humans…

If you go on the Internet, you can find many books and articles regarding Resilience Engineering (RE). It is important to understand that the RE considers technical systems (process, operation, activity, etc.) which can adjust its parameters before, during, and after happening of particular adverse events with intention to adapt itself to a new situation.

But, if we "transpose" the characteristics of the Resilience Engineering into the area of management, we can speak about Resilience Management (RM). If the RE applies to the Equipment, then the RM will apply to managing of the company.

As definition of RM, I can give my "contribution":

> *Resilience Management is methodology for skillful monitoring, controlling and guiding the "atmosphere" in the company with the intention to notice possible weaknesses in the behaviour of employees, to implement measures to make them enough strong and adaptable to all kind of situations (normal and abnormal) and to be able to provide fast recovery after abnormal situations happen.*

The application of the RM in the company's SMS should enhance the overall ability of the company to create procedures that are "tough" and "elastic" using the available resources. The adverse events in the company, considering the humans, are not result of faults of the system, but they are connected by non-capability of

[12] In some literature, resilient system and autonomous system are synonyms. I do believe that autonomous system is more advanced than resilient system and that is the reason behind the explanation in this paragraph.

the employees to adapt on the variations of the operations in the company caused by internal and external influences.

The Resilience Management, as a concept, is dedicated not only to take care to stop bad behavior but also to find suitable measures to support good behavior and to keep the good atmosphere active as long as it is possible. In general, we can look on the RM as a management system which can help companies to contribute to good human behavior during everyday operating activities by correcting the bad situations to minimize effects of failures and to maximize the effects of the normal operations.

Trying to build a management system which is "tough" and "elastic" means that it needs to be dynamically adaptable regarding his "toughness" and his "elasticity". For many of the readers, "tough" and "elastic" are two words which oppose each other, so I will offer additional clarification…

Every human is prone to stress during and beyond his working hours. In the scope of this book, I can mention that the humans try to cope with the most of the "known stresses" which happen during normal functioning. In the case when the company (during its normal operation) experiences some adverse event and the company does not change the "normality" of its operations, I can say that the company is "tough". Another case is if (during normal operation) the company experience some adverse event which is not catastrophic and it only affects the daily operation, then, if after eliminating the causes of the adverse event, the company recovers its normal operation immediately, then the company is "elastic".

So, from "the context of the things" in the RM, I can say that "tough" company is a strong company, where the working atmosphere is enough strong to oppose every interaction with the adverse events. Regarding the "elastic" company, I can say that it is company which transforms adverse events into something sterile inside itself and it can recover operation later.

We can easily create the "toughness" of the company in the cases when all these adverse events are known. We can embed counter measures to fight these adverse events. Regarding "elasticity" we can use it only with the cases with unknown adverse events (risks hidden in the P_{NA}). The problem here is that we may just assume (with unknown probability) the quantity and quality of unknown adverse events. That is a problem, because the "elasticity" of the company cannot be produced with the same certainty as "toughness" due to the IGs. So, in general, we can build "elasticity" in our company, but the uncertainty of how successful it will be will still be high.

In theory, as I have said previously, we can "fight" known adverse events by "toughness" embedded to our company and we can "fight" unknown adverse events by "elasticity" embedded in our company. But, in reality, there is need to make a compromise between the efforts for producing the working atmosphere (which is "tough" and "elastic" in the same time) and the overall organizational management of the company.

The role of the humans must be in the maintaining of the established RM. Whatever type of the RM is produced by the company, the employees have a duty and obligation to contribute to it. The RM can be designed and maintained by the managers who are experts in Human Factors and who understand the complexity and adaptation of humans as employees. Such a manager cannot be easily recognized, and this is a problem for Human Resources (HR)…

The main question with the RM is: Can we create a safe company if the employees who work there are not resilient? I do not know, but obviously there is need for something similar to RM which shall start to be implemented inside the HR departments also.

The human's resilience can be achieved by looking not only to experts but also to the employees which are open-minded, flexible, and creative and with particular ethical behavior. Obviously, only expertise (although is highly desirable) cannot be treated as most important requirement for the new staff, especially in the Risky Industries. There is need for particular personality which will fit in the company and incentivize, at the same time, the company's atmosphere of professionalism, dedication, excellence, expertise, and, of course, safety.

The RE and RM, together, support "the context of the things" of Safety which is not measured by the absence of adverse events, but it is measured by the presence of resources and capacity to handle adverse events. Adaptation, created by RE (for Equipment) and by RM (for Humans), means building an adaptive capability of the assets inside the company.

9.17 HUMAN FACTORS AND BEHAVIORAL SAFETY

In many places in this book, I have mentioned the Human Factors (HF), and they are a big factor in building the Resilience Management. Here, I will make connection between HF and, so-called, Behavioral Safety (BS). Both of them will contribute to implement effective and efficient RM.

But let's first explain something very important to understand the connection and difference between Human Factors (HF) and Behavioral Safety (BS).

As already elaborated in Section 4.2 (How We Calculate the Risk in the Risky Industries?), there are two types of safety which we deal with in our professional lives. First is Occupational Health and Safety (OHS), which is very often accompanied by Environmental Safety and it can be found under name Occupational Health, Safety and Environment (OHSE). The another one is the, so-called, Functional Safety.

The OHS and Functional Safety do not exclude each other. They, actually, complement each other depending on the place which you try to make safer. From the point of view of the root cause for adverse events, in the Functional Safety, the humans are the cause for approximately 70% of all adverse events; in the OHS, they contribute more: Approximately 88%.[13]

Taking into account the title of this paragraph, let's say here that HF is more implemented in the Functional Safety and BS is more present in the OHS. Let's try to explain in more detail the Human Factors and the Behavioral Safety in the following two sub-paragraphs.

[13] These values depend from the source, so they can be taken not so precisely as stated. Anyway, it is true that in the general industry, the humans are bigger root-cause for adverse events then humans in the Risky-Industries. Reason for this situation is simple: In the Risky Industries the employees are chosen by their knowledge, skills, experience and attitude and as such, they are more aware about possible bad consequences. So, they are more careful and more disciplined during company's operations. In addition, the monitoring and automation there is bigger, which also contribute to have less adverse events.

9.17.1 Human Factors

The Human Factors are outcomes of the scientific approach to the aspects of human behaviors that contribute to incidents and accidents. They mostly apply to the Functional Safety, especially in the Risky Industries, but it does not mean that they are excluded from the OHS. In the OHS area they are used also, but there is no so strong (direct) Regulatory requirement to implement HF program.[14]

I will not provide too much information regarding HF because it is a separate discipline in the Safety Psychology and there is plenty of literature dealing with this. But I will provide a simple table (Table 9.1) where the HF and some guidance for their management is described. Speaking about the "Dealing with HF" means speaking about the way how to identify them and how to eliminate and/or mitigate them.

There is another thing which is important to be understood and this is connected by the Invisible Gorilla and illusions which we have about ourselves. It is clear that in processing data and particular probabilities, we, as humans, are prone to overestimate the chances (events) for positive outcomes than those for negative outcomes. Underestimating the negative outcomes could contribute to decreasing our stress and it is beneficial to our well-being. And the same is why we overestimate the chances for positive outcomes. Anyway, have in mind that both things (underestimation and overestimation) have good and bad outcomes. So try not to exaggerate in both cases and try to find the right balance. This is a part of the Stress as Human Factor and, in general, having positive attitude, being optimistic, could help in dealing with any "safety animal". Of course, if you exaggerate, then optimism will transform itself into overconfidence, or even arrogance, and there is nothing good in that.

9.17.2 Behavioral Safety

If we can say that the HF use science to explain the reasons for human mistakes and errors, the Behavioral Safety[15] (BS) deals with wrong human behavior which could be actually triggered by the HF. To be more precise, the BS is actually a guide how to behave (encouraging good practices) not to produce the HF or adverse events and how not to be influenced by the adverse events (if the HF is present). The BS should be designed and implemented in such a way to discourage the bad practices and encourage the good practices. If an adverse event happens, the BS should show the way how to behave to minimize the damage. Actually, if implemented in the company, the BS could improve the safety by changing the behavior of the employees.

Having in mind that the adverse events do not happen very often, the BS causes a shift in the human behavior: Humans become more negligent as time passes believing that nothing bad could happen. This overconfidence in the Risky Industries is very dangerous and I have explained such consequences in many paragraphs in this book. The BS actually tries to "embed" into humans a good behavior which should,

---

[14] Actually, to be honest, nevertheless there are few standards about this, it very much depends from the States in different parts of the world. So, the statement above needs to be taken only in the general "context of the things".

[15] In literature you can find also name Behavioral Based Safety (BBS) and there is nothing wrong with that.

TABLE 9.1
Human Factors

Human Factors	Dealing with HF
Fatigue	There is a need to identify the reasons for the fatigue caused by events inside and outside the company. It is important to recognize the effects of fatigue on the worker's performance inside the company. Based on data gathered and processed, the company shall implement a fatigue relaxation program which will deal with the causes and the consequences.
Stress	There is a need to identify the reasons for the stress caused by events inside and outside the company. It is also required to recognize effects of the stress and to provide remedies for the stress and program for relaxation in the company. Maybe it looks funny, but having a place with boxing sack is good solution for the stress.
Alcohol and other drugs (AOD)	The effects of the AOD use are well identified as huge risk factors in the human performance and, as such, there is need to have zero tolerance insidc the company for the use of AOD. Anyway, there is a need to deal with prevention and to maintain particular periodic AOD testing.
Team cooperation and coordination	There is a need to have proper understanding of the Team structure and individual duties inside the Team. There is a need not to interfere with other's responsibilities. This is a way how the Team Management should function: Proper and on-time solving of personal conflicts, striving to exchange valuable information in real time, assuring better coordination, etc. All these things can help with this HF.
Decision Making	Gathering data with integrity from reliable sources, defining the cause of the problem, considering all possible available information and options for solving the problem, and consulting the history of activities connected with the problem could help to find unbiased decision on what to do.
Situational Awareness	Registering adverse event as soon as possible, spreading detailed and accurate information to the Emergency Team in timely manner, trying to understand the problem and all possible available options, choosing the proper one (based on situation and available resources for implementation) is the right approach.
Communication	Try to encourage sharing important information on a clear way and concisely. Try to include not only the context, but also the important details. Push yourself to listen carefully and think after that what was just said or communicated. Try to identify and address possible barriers to communication in the company and do it in advance. Prefer written interchange of information and data, because later you can prove what is communicated. There shall be a procedure for interchange of information and how to communicate inside and outside the company.
Leadership and Top Management	This is the obligation of the CEOs and managers. They must use their authority by maintaining the human standards. Wrong planning, wrong prioritizing of activities, and poor managing of the resources is very important HF that affect the employees. Having in mind their responsibilities and accountabilities, no one can make bigger damage to the company as it can be done by the CEOs and other managers.

actually, behave as Poka-Yoke in regards to their behavior. The employees should be trained on safe behavioral practices with the intention of HF not to influence their behavior. This is something used in military: The training of soldiers is so intense, so soldiers instinctively provide good reaction in dangerous situations.

The first BS measures were implemented in the late 1970s and there are few steps which are part of every BS program:

1. **Specification of the Type of Behavior Which Will Provide Particular Safety Performance of the Humans** – That is the reasons that, when we produce the Operational and System Procedures in our management system, the behavior of the humans must be taken into consideration for each process. Particular steps to maintain safe practices during the process shall be part of each procedure. During the training, explaining what can go wrong, if a procedure or particular step in the procedure is not followed, is obligatory. These explanations could prepare employees for the most of all possible adverse events;
2. **BS Must Be Connected with the Safety Performance** – This is actually part of the OHS and Functional Safety management systems, where we need to measure a safety performance as assessment of achieved goals. The BS's practices must be implemented in the procedures and appropriately assessed, in the scope of safety performance assessment. This is actually a merit of our BS practices in reality, especially in the areas where investigation showed that the root cause for some adverse event was inappropriate human behavior;
3. **BS Must Provide Methods for Feedback to the Employees** – The employees are those who work in the company and they actually execute procedures in the real time. As such, with their knowledge, skills, and experience, they are the first viewers of the problems with the operations and the procedures. So, any draft of new procedure should be submitted to the employees who will use it and their feedback must be analyzed with due attention. In addition, there must be channels which are capable immediately of providing feedback about operational and safety performance in the company to the managers;
4. **Managers Must Not Delay Assessment and Measures to Improve Operational and Safety Performance** – This is connected with the previous step: If there is negative feedback from employees, the responsible managers must immediately investigate the nature of the feedback. Having back-up plans or contingency plans could help in emergency situations, but the managers shall not delay their responses and they need to think in advance what to do, if the BS's practices are not good.

The problem with BS is that not everybody is assured that BS can really bring benefit to the OHS. There are many experts who think that it can bring problems. This is something well known in the safety management: Whatever solution you will find for your problem, it will introduce new or novel hazards. That is the reason that there is Regulatory requirement: every change in the system is to be subject of new (repeated) risk assessment.

From the negative opinions about the BS, I can say that many managers are not happy with the BS by stating that some of the BS programs are more focused on blaming of employees not taking care of working environment and on immediate symptoms than on the real root cause. I can agree with all these things, but in my humble opinion: It is a flaw caused by the badly created and poorly implemented BS program or by ignorant and bureaucratic management. The same thing can happen with implementation of programs for the HF in the Risky Industries, but there is something else which can prevent it. The Functional Safety, especially in the Risky Industries, is based on "Just Culture".

"Just Culture" is a concept implemented in the Functional Safety that provides more information about safety events gathered from employees by guaranteeing to the employees that they will not be punished if they report the safety events (regardless of their involvement).

In general, all of it (the HF and the BS programs) depend on the gap between the type of safety you would like to produce and the type of safety which is actually produced.

9.18 LIST OF HAZARDS AND UNCERTAINTY

There is no secret about the Risk Assessment in the Risky Industries.

The procedure how it will be done is well established and applies to each industry. There is a need to provide a List of Hazards where all possible hazards in the company will be determined through brainstorming session of employees with different background. It is good if the List of Hazards is communicated to any other employee. Maybe some of them would add a new or novel hazard or someone would comment to the already listed hazards. After that, the process of Risk Assessment starts: For each of these hazards, there is a need to determine the frequency (likelihood, probability) how often they will happen and severity of consequences (if they happen).

When this is finished, it is wise, again, for the determined frequencies and severities of the consequences to be communicated to other employees. Have in mind that these employees are in the "first lines of the battles", they are well aware about the processes. Any information or comment which they could provide in this phase will decrease the uncertainty about future events and about the measures how to prevent them. So, the submitted comments from them, shall not be neglected!

After that, the process of Risk Management could start. For each of these risks, there is need to find a way to eliminate or mitigate it. This could be done on two ways. The first one is to provide measures which will stop risk from happening (elimination) and/or provide the measures to decrease the frequency of happening (mitigation). The second way is to provide measures to eliminate or mitigate any consequence which will show up if the hazards materialize.

Having in mind the process of Risk Assessment, it can be noticed that the only known (deterministic) things in this process are hazards. The frequency of happening and the severity of consequences are actually unknown things and they vary from case to case. The variation of them is caused by the external or internal factors. So, in general, we can treat the hazards as something which is known and frequency and severity of the hazards can be treated as something which have to be determined.

In this job, we actually try to provide some prediction and, as such, we must understand that the determined frequency and severity are full of uncertainty. The job of the Safety Managers is to decrease this uncertainty, as much as it is possible.

With time, we will gather more data and more experience. We will be more familiar with the equipment (operations, processes, activities, etc.) in the company. All these things, together, will help to decrease the uncertainty, but having in mind that the Risky Industries are dynamic entities, there is need to keep, all the time, an "eye" on the data and its processing. In addition, the dynamic nature of the activities in the Risky Industries could also create, from time to time, new or novel hazards which are not in the previously prepared List of Hazards. So, this process of Risk Assessment is iterative and repeatable.

But, in this overall process of Risk Assessment and Risk Mitigation, something is hidden which many of the Safety Managers are not aware of. It is interesting that Ms. Wucker is addressing this thing in her book. Let's try to explain this from "the context of the things" in the Risky Industries…

You can notice that I have mentioned in this paragraph that during creation of the List of Hazards, we need a group of employees with different background. In addition, after the List of Hazards is created and after determining the frequencies and severities, it is wise to spread this information to all employees asking for comments. This is not just expression of democracy in the industry. This is very much important simply because different people with different culture (education, social status, religion, nationality, etc.) could provide better analysis without a bias, if they work together. This is something which is stated in the Ms. Wucker's book (The Gray Rhino) on the page 67 (Down with Groupthink).

By investigating the influence of the group-thinking to the individuals, Ms. Wucker has spoken to Mr. Frank Brown[16] who explained that actually, the groups of humans with manifold cultures, social status, religion or education of the humans, are not the issue. This is true simply because the researches have shown that these manifold groups provide better decision-making than the groups of humans with the same background. In addition, the discussions within these groups are more productive and provide more details which help the decision-making process. This comes from the fact that civilized and creative people have more respect to differences in culture, religion, social status, and education, than ordinary people, so flexibility of considering others' opinions and accepting the compromise is not an issue for them.

So, spread safety information to as many people as possible and build an atmosphere of commenting and sharing anyone's own opinion within the company. Again, it has nothing to do with democracy, but it has plenty to do with the effectiveness and the efficiency of your SMS. This is something which good Managers shall know!

The point of this paragraph is that I speak about a live system which changes dynamically and dealing with the uncertainty inside is not easy. We use statistics,

[16] Mr. Frank Brown is Managing Director of General Atlantic. He has extensive business and global risk assessment experience and he is frequent speaker on leadership conferences. He is author of The Global Business Leader: Practical Advice for Success in a Transcultural Marketplace.

probability, technology, human factors, and everything else which, we assume, can help with the uncertainty. In these efforts to make the things better, the knowledge is very much important and I hope this book will help with increasing your knowledge. Do not forget: Our limitation of knowledge applies also to the type of hazards which can show up in our activities. But this is not a problem with Risk Management. The real problem is to determine the frequency and severity and, there, our knowledge is masked with probability and it is full with uncertainty.

9.19 WHAT TO DO IF THERE ARE FEW "SAFETY ANIMALS" AT THE SAME TIME?

Whatever I have written in this chapter deals with the measures what to do with the "safety animals", if any of them show up.

But, what to do if there are few "safety animals" present at the same time?

What to do if, in the very short period of time, few of them just pop-up?

Or one "safety animal" transforms itself into another?

As mentioned previously, the impact of this "safety animals" can be catastrophic, so having one or few together in the same time can create only bigger catastrophe. Also, for those who read the book carefully, they could notice that, as in the nature, these "safety animals" are connected even in the area of Safety and they can affect or transform themselves into others. That is the reason that, whenever, during monitoring, the things start to look irregular or abnormal, there is a need to react immediately. Of utmost importance in these situations is to recognize and define the true hazards and risks immediately.

The first thing is to try to understand what is going on. For this to be effective, the employees need to be educated and to have experience. You must understand that in the case of two (or more) "safety animals", the available data is coming from two (or more) sets. In such a situation, no accurate processing is possible and no probability distribution can help. In the first moments, the quantity and the quality of the data available are not so big, so the employees must check all data and, if unconclusive, look for more data. If, in this stage, there is no hint what is going on, then the switching off the system (landing of the aircraft, switching off the nuclear reactor, stopping the chemical process, etc.) is the best solution.

To be honest, many of you will say that it will be timely and cost-consuming to recreate the previous situation later, but for those who have doubts is this really expensive, check the Deepwater Horizon example on Internet. The management of British Petroleum during the closing of the well did not follow the procedure. They received unconclusive results from testing and they did not know what was the meanings of the test results. So, in their wish to save 200,000 USD per day (just for few days), later, they have to pay approximately 65 billion of USD for cleanup, charges, and fines. If you think that the safety is expensive, try to calculate the cost of the adverse events after they happen!

If anyway, the employees could recognize the "safety animals" which are on its way to happen, then they must decide in the moment which one of those two (or more) will be handled first. The criterion for which one to choose is very clear: The one with the worst consequences shall be handled as soon as possible.

The second thing to do is to try to contain the event. Fire is good example of that. Contained fire will be beneficial for the human lives and also will be beneficial in regards of the damage of the assets. In this stage, it is important to execute the emergency procedures, the contingency plans, and the back-up plans. All available resources shall be used, and particular systems for emergency communication, synchronization, and cooperation must be established immediately.

The third thing is to deal with the adverse event by all available resources.

The fourth thing is to try to contain (eliminate or mitigate) the consequences. Timely information to the police, to the fire services, and to the medical staff could make a difference between incident and accident. In many states, it is obligatory to have Search and Rescue services, and they must be contacted as soon as possible.

The fifth thing is to try to preserve the site where the adverse event just happened, so the investigators from the police and the State Regulatory Bodies can try to find the evidence of how and why it happens. The primary reason for that is to "learn a lesson" so that the same thing does not happen again. If great negligence is discovered, then, the police would join the investigation.

The sixth thing is to publish report to the public, so everyone will know what, why, and how it happened. As said previously, this is a Regulatory requirement in the Risky Industries and the information hidden in the report must be spread to the public.

The seventh thing which needs to be done concerns the Regulatory Bodies in each State: They shall implement the directives and recommendation given in the report in their industries.

This is the way how things shall be done...

Another thing which needs to be considered is the answer of the question: What if we are expecting BSe and the event is actually GRe...?

Looking for answers, many times, we are in a situation masked by any of the illusions of Invisible Gorilla (IG). The mistake in deciding what is going on is highly probable simply because the data are not abundant and each man has his own individual bias in providing the answers. Of course, this is not good, but it is not so critical. Following the procedure explained above could help in dealing with such a situation.

10 Top Management and "Safety Animals"

10.1 INTRODUCTION

The most significant safety (and quality) issues inside the companies associated with the bad management are:

a. Lack of training for employees for executing the operational and the system procedures;
b. Saving money by decreasing the number of staff (understaffing). This issue has two other bad consequences: Increases the complexity of the activity which needs to be done by one person and increase the stress to this person;
c. Pressure by production schedule, by unrealistic planning, or by neglected influence of some other events (internal or external) on the production plans;
d. Bad, missed, or delayed preventive maintenance and bad, delayed, and/or improvised corrective maintenance due to lack of spare parts;
e. Habit for blaming the employees for delays in the production or for the adverse events;
f. Complexity of the jobs and requests for non-adequate education, experience, and skills of the person dedicated for such a complex job (trying to save money by employing under-educated persons);
g. Not considering environmental factor's influences on the production processes (heat waves, hurricanes, snow storms, floods, etc.);
h. Neglecting already identified hazards and calculated risks;
i. Not understanding Human Factors in total! If you do not understand something, you cannot fix it;
j. Trying to fulfil bureaucracy instead of producing useful measures and actions (there is everything on the paper, but nothing functions in the reality!);
k. Etc.

Today, the world and almost every aspect of our lives are very much complex. The complexity is also included in the industry, because the new and novel technologies need shifting in our attitude, methods, and tools to maintain the production or service-offering processes.

As it was mentioned in Section 1.3 (Determinism, Randomness, and Chaos), the complexity produces uncertainty, and this uncertainty is one of the reasons for unpredictability. Today's way of doing business is mostly based on known things and implementation of very much well-known methods and "laws," which helps the Top Managers to make decisions regarding the strategy and tactics of their businesses.

DOI: 10.1201/9781003230298-10

Today, the increasing complexity actually makes this process harder to maintain and, obviously, there is a need to change something. The new technologies, new products, and services require new ways of approaching and solving the business problems.

No more is there just one person who is in charge of everything, but there are persons who are managers of departments and each of them is in charge with particular area. In other words, the complexity needs a Team to handle all these things. So, there are no Top Managers anymore, but there is a Top Management. Nevertheless, in reality there are still guys named Chief Executing Officers – CEOs (or General Managers, General Directors, etc.), who can be entitled as Top Managers. Their decisions are made on the basis of team meetings, interchange of data with other managers, and opinions and common understandings of the problems in the company. Strategic company decisions are produced in synergy with management teams (Top Management), which are usually made up by the Top Managers and the managers of departments.

One of the preconditions to be successful in any fight against the any "safety animal" is to have established clear "chain of command". This is something which can be determined by the Top Managers. In the good companies, this is already achieved and any process can be executed very effectively and very efficiently. In the bad companies, this will actually add "fuel to the fire", so the things, hardly, will not go in good direction.

10.2 MANAGEMENT TEAMS

We need a Team in the industry. These are different types of the teams, and they need to be established because the volume of operations and processes in the company are complex and they cannot be done by one person. It means that each team consists of employees with different expertise and they contribute with their expertise to the common goal.

A good example of the use of team is medicine. There, the diagnostics of your medical problem could be established by one doctor and solving the problem could be done by another doctor (surgeon, for example). The main point is that one doctor should be knowledgeable and experienced to provide diagnosis and another one should be knowledgeable, skilled, and experienced to solve the problem. In "the context of the things:" in the Risk Management, the "safety animal" can be noticed and determined by one employee and the few different employees (a maintenance or emergency team, maybe) could deal with the adverse event.

How I can explain a Team?

A Team is not just a bunch of guys who are doing the same job. A Team is bunch of guys and each (or few) of them is dedicated to particular different operations (area, activity, etc.). Metaphorically speaking, the teams are actually machines with many parts and each of these parts is doing a particular job.

The philharmonic orchestra is a Team…

The members of philharmonic orchestra are musicians, and most of them play different instruments. All together, they play complex musical pieces which (maybe) can be played by one instrument, but their beauty is hidden in the simultaneous performance of many instruments. In the philharmonic orchestra, if the oboist is sick, he cannot be changed by violinists. Actually, it can happen, but the orchestra is not anymore an orchestra: It is an improvisation.

The departments in the company are teams and the company, itself, is a team consisting from departments. So, in the scope of the changes of musicians in philharmonic orchestra, if transferred in area of industry, I can say that, the interchanges of the managers of departments could not happen very often. Actually, it happens very often because there is a lack of understanding what the Team is. You, simply, cannot put an economist to deal with maintenance of the equipment in the factories.

OK, OK... I agree...

You can put one in, but let's be honest: It is like putting an economist in the hospital to do a medical examination of patients. Would you be happy for your child (or even you) to be examined for medical problem by the economist???

So, the attitude that a manager (part of the Top Management), could help the Team by undertaking the job of other manager (who is not available due to any reason), is wrong. That is the reasons that each manager must have deputy manager. This is a guy in the "shadows", who is aware about the situation in the department and he can undertake the job if something happens to the main manager of the department.

Once on LinkedIn I saw a post (a video) showing how the military recruits help each other in overcoming high wall (put as obstacle) during their every day's exercise and it was mentioned that this is great example of Team Work. I express my disagreement of this that helping each other is not Team Work and I mentioned example of philharmonic orchestra. From maybe 40 comments, only one guy supported me. All others did not provide any comment. Using Bayesian statistics on this rudimentary example, it means that only 2.5% of the employees in the industry understand what is a Team Work.

10.3 RESPONSIBILITY AND ACCOUNTABILITY

Going back to the management teams, not all of the managers inside the Team are equal. There must be one guy ("the captain of the Team") who listen to all "pros and cons" and he must make the decision. It is the CEO (Director General, President, etc.). In the scope of the Management Team, all managers are responsible for the things in their areas of expertise (their departments) and the CEO is accountable for all company. So, the CEO's responsibility in the decision-making processes is transferred to accountability for the situation in the company. He is the "guy with the money", because his signature must be on all payment documents. So, imagine a situation where the Safety Manager propose some measure which can help with some risk and the CEO says "It is too expensive, we will not buy it". If this risk will materialize as incident or accident tomorrow, the blame will be on the CEO.

The accountabilities of the CEOs, as final decision-makers, are not easy. Most of them, in the presence of lack of good information and associated uncertainty, very often decide not to do anything. Doing nothing, sometimes, is a wise decision. But this is gambling: If the adverse event does not happen everybody is happy, but if it happens as result of the wrong decision or no decision, the price which needs to be paid is terrible! It is clear that, in such a situation, the CEO will lose his job and, maybe, go to prison due to negligence.

The accountability of the CEO can be seen mostly when the bad things happen. Their first response (almost always!) is to limit the public relation's damage by restricting the flow of information with the outside world. It is clear that it will not help company to solve the issue with the adverse event, but it could save the face of the company in public. At least, they think it will happen…

10.4 TOP MANAGERS

Today, the Top Management can be also associated with the Boards in big companies, where the members could be stakeholders, some independent experts, academics, or even politicians. I do not like this association, because there is a difference between the Top Management and the Boards in the companies: The Boards usually deal with strategy of the companies and Top Management deals with tactics of how to implement the strategy. Having in mind that the Boards deal also with the finance and the money, then their strategy can, very much, determine the tactics.

Good CEOs (or other Top Managers) will direct companies to create value through entrepreneurism, innovation, and research and development. They will try to provide accountability and responsibility to the management decisions commensurate with the risks involved. And they will try to provide suitable real-time monitoring and control systems to see where the company is going. They deal with what will happen today, but they must think what will go on tomorrow or next week (month, year, etc.). Dealing locally and thinking globally is something which good Top Managers must do.

In addition, they will support and encourage the involvement of every employee in providing comments, advice, or opinions about things which are important for the production process or regarding the quality of the products and safety. These things can come from inside and outside the company. A good manager would understand what is written in Section 9.18 (List of Hazards and Uncertainty) regarding group-thinking and the benefits which it can bring to the company.

The bad CEOs (or other Top Managers) could guide the company in many different directions, but in the Safety area, usually a gap between what they say about the risks and what they do is most critical. The most important questions could be: Does the CEO (Top Manager) provide appropriate behavior which can be used as a criterion regarding the risk-aware culture and the safety policy which is on the walls of the company? Does his behavior encourage "just culture" thinking and behavior? Does he support preventive and corrective actions as planned to be executed?

I am very much aware about Top Manager's ignorance in the aviation, and I do believe that this ignorance is present also inside other Risky Industry's companies. I am not saying that there is an ignorance in the expertise regarding the used technology, but there is ignorance regarding the management of humans.

One of the biggest ignorance shown by the managers is that most of them are proven in their expertise with the equipment and the processes in the company. But the issue is that the Top Managers do not deal with the equipment, they deal with the Humans (employees!). If we speak about the influence of the Top Management to the Equipment, we speak about engineering, not about management. The Top Managers should not manage the Equipment; they should manage the Humans.

This ignorance is pretty much present in the Boards of big companies where, not always, the members of the Boards can recognize the operational risks. This failing to recognize operational risks from the members of the company's Boards contribute to a Cognitive Bias. A Cognitive Bias is a systematic error embedded into the human behavior that affects the judgments about the root causes of the problems and the decisions that people make (based on that judgment).

Another aspect that puts the Top Managers in situation to "help" the incident and accidents to happen is that they must dedicate themselves to earn the profit for the company. As it has been explained in Section 6.4.1 (Defense Lines of the Company), by dealing with the activities to provide a profit for the company, they can neglect the duties to protect the company, humans, assets, and environment. So, in the Risky Industries, with the intention to have control over costs dedicated to safety, the ALARP principle is used. The problem is that its use is very much neglected by the managers. The Top Management is prone to put the risk in their operations, if it can bring bigger profit to the company. This is something well known in the Stock Exchange and financial world, but sometimes it can be present also in other industries. Do not let me be misunderstood: I am strongly against those things, but in the reality, these things happen.

This problem is emphasized even in the areas of emergency procedures, back-up plans, and contingency plans, because their effectiveness and efficiency can be only assumed and it cannot be seen beneficial in the efforts for bigger profit. That is the reasons that many Top Managers use the "cuts" in these areas to save money. These are procedures and plans which need to be activated immediately when the bad things happen and, as such, they are a matter of death and life. When the bad things happen, these plans are actually decreasing the damage and losses for the company and there must not be any compromise with them: All possible resources should be available for these plans in such situations!

There is one statement of Todd Conklin who is a well-known expert in the area of human performance and safety integration. He has said that incidents/accidents are not caused by employees as it seems in the most of the cases. These are caused by the latent conditions in the companies and the incidents/accidents lie dormant and hidden inside the companies. They just show up when the conditions are good for them. These are like viruses: They can stay hidden in terrible conditions for hundreds of years and, when the conditions are suitable, they will revive.

In general, the risk analysis theory states that most of the adverse events come from one or more of four levels of the management failures:

1. Bad company organization;
2. Neglected oversight of processes and operations inside the company;
3. Hidden preconditions for adverse events; and
4. The adverse events themselves, caused by bad management.

All these levels, eventually, are accountability and responsibility of Top Management.

It means that the accountabilities and responsibilities of Top Management are not only to provide profits to the company but also to create atmosphere of awareness, professionalism, and expertise, which will suppress the latent incidents/accidents in the company.

Simply I am prone to believe that most of the incidents and accidents come from ignorant Top Management and the atmosphere of bureaucracy which they create in the companies, especially, in the Risky Industries. I know that in aviation, 70% of all incidents and accidents are caused by human error or human mistake. Of course, it does not mean that equipment cannot fail. The equipment can fail (very rare, of course), but there are procedures and back-up and contingency plans in place, which are used to eliminate or to mitigate the equipment fault, so the operations can continue. Unfortunately, this is the place where the humans fail also: They fail with the use of these procedures and the back-up and contingency plans in critical situations.

10.5 UNDERTAKING RISK AND TOP MANAGERS

One of the reasons that the companies are "attacked" by "safety animals" is connected with the fact that many CEOs undertake risks when there is lack of information or there is high uncertainty of the available information. Not always do the things develop as we would like. I realized that the humans should establish clear rules when it is worth to undertake the risks and I try to implement these rules in my private life. There are three such rules:

1. Never undertake a risk if the damage is bigger than the benefit;
2. Undertake a risk only if you are sure that you can live with the consequences; and
3. It is wise to think twice, even more times, before you decide to undertake a risk.

The humans usually transfer their individual principles to the society, and this happens also with the managers. That is the reason that the Top Management should be morally fit to deal with the companies. Do not forget that the Top Management takes care not only for themselves or the company but also for thousands of families of the employees which can be endangered by their risk undertakings.

I have already presented the facts about these misunderstandings in few of my papers in conferences and symposia and I am trying to "spread word" during my teaching activities. But (I must confess!), I am not so successful in these actions to emphasize the importance of continuous education and keeping the pace with the newest methods and technologies in aviation. I could notice that most of the working people are in "hibernation" regarding their knowledge and when they reach some particular advanced position inside the company, they simply neglect their own improvement and they stop to follow what is going on in the industry.

And this is a good terrain for the "safety animals"!

If you are not in touch with the new and novel ideas and events in your industry, you will not follow the movements and dynamics on the market and "the bad things happen to the good people". In such a case, it is reasonable if your customers will choose another supplier for the same products or services, instead of your company.

When I was appointed to one managerial position which had to be approved as Post-Holder from a particular State Regulator, I had to attend one-day training specially shaped for Post-Holders. Together with me were eighteen other new-appointed managers and during the training, the instructor tried to explain some basics regarding the Safety Policy. Plenty of the managers were appointed in operations area and most of them were not familiar with the SMS principles. I found it as very big paradox: You will be Operational Manager in the Risky Industry, but you are not familiar with Safety Principles (?). That was my understanding of this one-day training.

The main point was that the instructor's explanations had nothing common with that what the Safety Policy should be. And I (frustrated) reacted to that. To be more specific, the instructor stated that Safety Policy is not so important for the SMS and for employees (?). When I reacted by saying that is not true, in the next moment, 6–7 other attendees started to defend the instructor. One of them said: Why should a simple technician in the company know what is the Safety Policy? On my question, in his company, is the Safety Policy written on the walls, he responded in the affirmative. But, to my question why, if it is not important, it could be found on the walls, no one of the present managers had an answer…

You may imagine: I am speaking about Risky Industry's managers, but not one of them knows that Safety Policy is a "Constitution" of each SMS. Not one of them knew that you need the Safety Policy as a guide of what to do if something happens and this something is not covered by any procedure. It means that, in such situations, you must improvise, but you must do it in accordance with your company's Safety Policy. If you do not know the Safety Policy, you are in trouble.

As a warning to the managers, I can say only one thing: In the history of aviation, there is no small airline which has experienced accident and has survived. All of them do not exist anymore. And the Top Managers (CEOs, DGs, etc.) must not forget that: Maybe the Safety Managers are responsible, but the Top Managers are accountable!

10.6 IG AND MANAGERS

There is one aspect regarding the Invisible Gorilla that usually spoils the effectiveness and efficiency of the managers. Actually, it affects very much overall performance and working atmosphere in the companies and it is very critical, especially when it will happen in the Risky Industries. It is called: A Bad Manager!

You can find bad managers everywhere. These are guys which are Post-Holders of managerial positions, but they do not have capabilities to manage their duties. Actually, when we speak about the managers, we speak about humans who need to deal (manage) other humans. It is very wrong when you put a good engineer (doctor, economist, etc.) just because it is good in his area to be a manager. Having good understanding or be specialist (expert) in your profession does not make you good with the humans.

One of my professors at university used to say that "manager is not a chicken and he cannot lay an egg, but the manager must be able to determine which egg is good and which one is bad". It means that being a good or an expert in your area of interest (engineering, medicine, chemistry, etc.) will not make you a good manager. The Management is about the people! In the management, you do not manage engineering equipment or medical devices or chemical processes: You manage humans!

So, how does the IG affect the managers?

I have extensive professional experience in different areas of industry, aviation, and education, which was also extended to different countries. I had met many people with different races, religions, and cultures. Many of them were worth of respect and many of them were not so worthy. I had a chance to meet many bad managers and few of them were actually reasons to change my employment status in their companies. To be honest, I have met more bad managers than good ones...

What a sad professional life I had...

All of these bad managers had few same characteristics:

a. All of them knew that they are not experts in the areas of their departments, but all of them were very much frustrated by that. When you try to explain them how the things are required by the regulation or by the laws of engineering, all of them were highly reluctant to listen or even think about what you were speaking;
b. All of them paid so much attention to their ignorance and they did everything others not to notice their deficiencies, so they could not notice the IG which was in front of them. They were focused so much of hiding their incompetence and ignorance, so not only they did not solve very easy problems, but they actually made them worse. I may need to mention here that they were also very successful in creating new ones;
c. All of them were so capable as manipulators, so you could not adore them because of that! I am still not sure: Do you need to have talent to manipulate humans or this is a skill which you can learn?;
d. Almost all of them had no understanding what are the Human Factors and what is their role as bad managers in creating the Human Factors, but they tried to providing some general document (manual) regarding "implementation" of Human Factor's remedies which was of no use;
e. All of them, when involved in the selection process for new employees, were choosing incompetent and ignorant candidates (managers, engineers, and technicians) simply because they were afraid that some of these guys could undertake their positions;
f. Etc.

In general, all these managers did not contribute to the well-being of the companies and all of them were forced to leave. Unfortunately, they were not replaced by better guys. The point is that when the football team is losing, you do not change the players, you change the manager.

As I mentioned previously, by the regulation in the Risky Industries, there is so called Post-Holder positions which are assumed as critical positions in the companies and you need approval from the Regulatory Body to assign particular person on that position. The reason for this position is a chance for the Regulatory Body to register the bad managers and it can reverse their selection on such positions. But, as it can be seen by my Post-Holder example from the previous paragraph, it usually does not work. Simply, you can find bad managers even in the Regulatory Bodies.

In my humble opinion, the bad managers have contributed to more accidents and incidents than any other human errors or mistakes. Of course, they did it not directly, but indirectly, by creating a bad atmosphere in the companies and by neglecting the obvious Invisible Gorillas.

10.7 CYNEFIN FRAMEWORK

Around the beginning of this century, a team of contributors, where the most prominent was David Snowden (an IBM researcher), established the so-called Cynefin[1] Framework. Having in mind the fast-changing dynamics of life today, this is a framework which can be used as a tool for dealing the with complex and unpredictable situations in the business and industry.

The Cynefin Framework (CF) takes into consideration multiple different factors in our lives and our experiences, which usually have an influence on us in a way we can hardly understand. The CF is used to better understand the dynamics of the new situations, the perspectives based on these situations, the possible conflicts, and requested changes in our minds to provide decision what to do.

The CF considers the issues by facing the decision-makings of the Top Management in regards to five areas, established by taking into consideration the relationship between the cause and the consequences (Figure 10.1). Four of these areas (Simple, Complicated, Complex, and Chaotic) are connected by diagnosing situations of interest of the company and they allow the Top Management to categorize each situation in particular area. The fifth one (Disorder) is used when the particular situation cannot be categorized into any of the previous four.

Two of the areas are connected with the certainty (Areas of Certainty), which means that there, in these areas, we know which situations we know ("Simple") and which we do not know ("Complicated"). The "Simple" and "Complicated" are marked with gray in Figure 10.1.

There are also two other areas which are connected with the uncertainty (Areas of Uncertainty). These are "Complex" and "Chaotic" (marked with white on Figure 10.1).

In addition, in Figure 10.1, the possible transition between categories are shown by arrows. In today's dynamic world, it is not strange that some situation develops in such a way to jump from one category to another. It means that, in the scope of the framework, it is important to follow the dynamics of any situation

"Simple" is a category of situations that we truly understand and we know how to handle them.[2] Using the language of NNT[3] and Donald Rumsfeld, we can call these situations "known knowns". This is the realm of determinism which "rules" this category.

In this category, everyone understands what is going on with the situation: It is stable, ordered, and predictable. Both the cause and the consequences are clearly recognizable in this situation: Everybody in the company agrees about connections between the causes and consequences. This situation abides by the well-known rules and uncertainty is very low here. This is the area where decision-making can be

[1] Cynefin is a word for "habitat" on Welsh language.

[2] Actually, there are more articles on Internet where you can find term "Obvious" instead "Simple". I noticed that the creator of Cynefin Framework is using word "Simple", so I have decided to use it here.

[3] See the Section 5.4 (Differences between BSe and GRe)

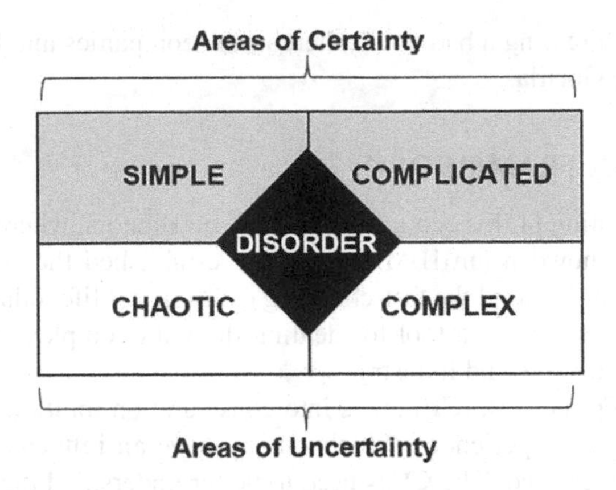

FIGURE 10.1 Cynefin Framework.

established simply by following the rules of Good Manufacturing Practice[4] (GMP) and, as such, it should not be a problem for any employee or any manager. The BEe belongs here.

"Complicated" is a category which can be associated by NNT and Donald Rumsfeld as "known unknowns". This is actually a situation which is included into our List of Hazards, and the associated Risk Assessment for each of the hazards is produced. Anyway, the Risk Management for each of these hazards is not provided simply because the relation between cause and consequence is not so clear. In this category belongs the situations which we have assumed that could happen and we have already knowledge (procedure, tools, back-up plans, etc.) about them, but we do not have experience and knowledge when and where they will happen. Obviously, we need help from the experts to handle it.

In this area, the cause and consequence cannot be easily connected, so there is need to establish different scenarios how these situations could develop. There is a need for additional investigation and for additional tolls and methods to establish these scenarios and to produce emergency procedures, back-up plans, and contingency plans to handle them. This is the area where the knowledge and experience are very much important and consultation with some experts is a wise step. Sometimes, maybe few solutions are available and there is a need for managerial choice what to do in the presence of such an uncertainty. For the situations in this area, the guidance of the Safety or the Quality Manager is necessary to help employees to cope with the problems. The GSe and the GRe belong here.

"Complex" area is a category which can be associated by NNT and Donald Rumsfeld as "unknown knowns". In these situations, we are aware about some aspects related to the problem, but they are not well understood. In addition, just small pieces of information are available about the problem and we need more investigation. Usually, after investigation, the answers are very much reasonable, but only

[4] GMP is acronym which somewhere is used for Good Management Practice and somewhere for Good Manufacturing Practice.

in hindsight. To deal with situations which belongs in this category, we need to prepare for the consequences. Prediction is not possible due to high uncertainty and maybe randomness of the situations. The NsNhNs, the DKe, and the OiS belong here.

"Chaotic" is a category which can be associated with "unknown unknowns". This is an area of industry where the Chaos can apply, but in addition, it is characteristic for the companies which are dealing with research of the new or novel technologies. Good knowledge and considerable experience are very important here, but they can be only partially used. The way of functioning of these companies is mostly experimentation, modeling, and simulation. In hindsight, after the experiment is finished, everything could become understandable, but until you reach a hindsight, the uncertainty is very high. This is the area where the BSe and the IG belong or, speaking by probability terms, it is the area of $\mathbf{P}_{NA}$.

The point with this categorization is that the complexity of the situation cannot provide a clear picture when a particular situation can be assigned to a particular category. For example, there is no situation which 100% belongs to particular category. It can be 60% into Complex, 20% into Complicated, 15% into Chaotic, and 5% into Simple. In this case, the situation will be categorized as Complex. If there is no clear understanding where the situation should be categorized, let's say, Complex -54% and Complicated -46%, then the situation belongs to Disorder. This is because Complex belongs to Area of Uncertainty and Complicated belongs to Area of Certainty, so these are pretty much different "context of the things". As it can be noticed, categorization is critical to provide good decision-making process. Making a mistake here could be very devastating.

Another very important note is that the transition from "Simple" to "Chaotic" can be sudden and also very devastating. Some of the authors are using metaphor for this transition as "cliff between the earth and the sea". The border (limit), when the regular (simple) process can become chaotic, is not well known, so sometimes "a quiet wind can become suddenly a tornado". This could happen very often if in the company there is tendency to over-simplify the causes and consequences. In such companies, the level of self-confidence is very high, but there is no good reason, based on the results, for such a thing. This is the reason that I will emphasize (again) the importance of 24/7 real-time monitoring, especially in the Risky Industries.

The managers, who are aware that the life is random and uncertain, and as such, it is unpredictable, will find benefit by using the Cynefin Framework. But there is very important thing which must be understood: The Cynefin is just a tool. Two different managers may put the same situation into deferent Cynefin areas based on their understanding of the reality, their knowledge, and their experience. So, as a tool, it is very useful, but there are limitations which very much resemble those seen with the modelling.

In general, there are many tools which are prepared to help managers in their decision-making processes, but the danger of the use of all these tools is expressed through the adage: "If you intend to use hammer, everything around you shall be treated as a nail". It is very much important to have good understanding what are the benefits from the tool and what are the deficiencies and the limitations of the tool.

Final Words...

So, this is a book about the adverse events (incidents and accidents), presented as metaphors which are used to describe different characteristics of these adverse events. These metaphors can happen in our lives, but the emphasis on each of them in this book is given by "the context of the things" in the Risky Industries. Everything regarding this book started with the book *Black Swan: The Impact of Highly Improbable* by Nassim Nicholas Taleb (NNT). The book become very popular, in that it gave a different "context of the things" on the class of adverse events that can happen.

The characteristics of the Black Swan events define a group of adverse events, but there are plenty of other adverse events which were not covered by this definition given by NNT. Anyway, I must say that the "fashion" to use the characteristics of animals as metaphors for the adverse events was accepted by other authors, and so today we have plenty of "safety animals".

A good Safety Management System, which can provide a holistic approach to any adverse event, should be able to deal with the incidents/accidents or, at least, deal with the consequences. Eventually, this is a book regarding complexity of the systems (processes, operations, activities, etc.) which must be considered to analyze some event from "the context of things" in the Safety and Risk Management. By increasing the awareness about what can go wrong with these "safety animals", we can produce a better understanding of the adverse events. Better understanding should bring better results!

Many of these "safety animals", unfortunately, result in catastrophic consequences, and it is very sad if we have to deal with them in hindsight. Another unfortunate fact is that the historical experience of humanity teaches us that we "learn" and make significant changes only after tragedies and pogroms and never as a result of benevolent opinions of experts about what would be better or what would be more convenient to do. Mankind "learns" like bad students from their worst mistakes that repeats along the way. Global Warming is an excellent example of that!

The Safety, as anything else in our lives, must evolve and must progress. It cannot stay in the same place and it must follow the development of new technologies by adapting itself to all these things. Whatever you are thinking about this book, I do believe that I have presented and emphasized here many new, forgotten, or neglected aspects of the Safety Management, and I do believe that this book can provide benefit to Safety Managers in their contribution to improve safety.

All these adverse events hidden behind the "safety animals" have happened previously, so the only things which are new are the metaphors which are given to them as names. But there is something more...

The classification by giving them metaphoric names contribute to increased curiosity of these events which contributed to considerable research in any of these areas. So, this is a new aspect to all these events. Introducing "the context of the things" for the Risky Industries to all these "safety animals" could point to better understanding of how to handle them in advance or later, when they happen.

Although there is considerable analysis of all these events, the reader can notice that there is no special recipe of how to handle them. In addition, the biggest problem with them is that many of them could be explained in hindsight, and this is not good. There is one adage which says: "After the battle, everyone can be a good general." Knowing what happened will help you to find a reason, but "hindsight will not make you a King". In the safety area, especially in the Risky Industries, what matters more is insight.

In general, we try to make ourselves, our families, and our friends safe. And we put a lot of effort into that, but the point of safety provided by us is that we just feel safe. It does not necessarily means that we are safe. There is a beautiful article[1] from Prof. Pasquale Cirillo named "Of risk, fences and unavoidable falls" where he explains something called Fence paradox. I would not explain here the Fence paradox, but I strongly recommend the reader to find this article (4 pages) on the Internet and read it. The main point of the Fence paradox is that we can feel safe but it is not necessary that we are safe.

From the point of the Safety Management System in the Risky Industry, its job is to provide safety, which can result with the humans feeling safe, but there must be other evidences (scientific maybe?) which need to contribute to "feeling safe". That is the reason that we use science (statistics, probability, technology, etc.) with the intention to contribute to the everyday safety.

I do believe that in the scope of the efforts to provide more safety to ourselves, our families, and our friends, we can do it by increasing our knowledge about us and the world around us. I also do believe that this book has pointed to a few very important things which will contribute to the improvement of the Safety in the Risky Industries. Now, when this book provides more details of these "safety animals", it is understandable that we cannot use the same strategies without reconsidering their relevance when some of these adverse events happen.

I hope someone will recognize it and, if this happens, it will make me happy!

Sasho Andonov
14/03/2021

---

[1] http://www.pasqualecirillo.eu/resources/fence_paradox.pdf (last time opened on 02/02/2021).

Index

Printed in the United States
by Baker & Taylor Publisher Services

Printed in the United States
by Baker & Taylor Publisher Services